Ernst Schering Research Foundation Workshop 10
Nongenotoxic Carcinogenesis

Ernst Schering Research Foundation Workshop

Editors: Günter Stock
          Ursula-F. Habenicht

Ernst Schering Research Foundation
Workshop 10

Nongenotoxic Carcinogenesis

A. Cockburn, L. Smith
Editors

With 24 Figures

Springer-Verlag Berlin Heidelberg GmbH

ISBN 978-3-662-03024-0   ISBN 978-3-662-03022-6 (eBook)
DOI 10.1007/978-3-662-03022-6

© Springer-Verlag Berlin Heidelberg 1994
Originally published by Springer-Verlag Berlin Heidelberg New York in 1994
Softcover reprint of the hardcover 1st edition 1994

Typesetting: Data conversion by Springer-Verlag

21/3130–5 4 3 2 1 0 – Printed on acid-free paper

Preface

"What is a nongenotoxic carcinogen?" This question recurred throughout the Ernst Schering Research Foundation Workshop on nongenotoxic carcinogenesis, underlining the complexity of the topic. The clarity of the view that all carcinogens act by mutating DNA, originally advocated by Bruce Ames nearly 20 years ago, has been clouded by the increasing numbers of compounds which are not genotoxic but which nevertheless can cause cancer. There is an urgent need to increase our understanding of these compounds so that their risks can be evaluated realistically and decisions made from a position of knowledge and strength, rather than in fear of the unknown.

A nongenotoxic carcinogen can be defined as a compound which causes cancer, but which does not cause damage to DNA as its primary biological activity. This negative definition covers a range of carcinogens acting through a variety of mechanisms. Such chemicals often produce tumours only in a single organ species, and there are a few common locations which are affected most often. For example, in male rats, certain carcinogens bind to $\alpha_{2\mu}$ globulin to form a complex which accumulates in the kidney tubular cells, which is followed by necrosis and compensatory cell proliferation leading the neoplasia. Other common mechanisms include hormonal imbalance resulting in thyroid tumours or peroxisome proliferation resulting in liver cancer. These and other examples are studied in some detail in the papers of this book.

Cancer is now widely thought to be a multistage process, and some of the stages of this process are examined, from DNA damage by the free radicals, through the progressive genetic and biochemical changes involved in the cells, to the influence of control mechanisms such as growth factors and tumour suppressor genes, the interactions of genes with hormones, and the importance of inherited genetic susceptibility to certain cancers. A prime reason for studying nongenotoxic carcinogens is to enable sensible decisions to be made regarding their risks for humans. The majority of known human carcinogens are genotoxic, and it is relatively straightforward for this class of carcinogen to take steps to ensure human safety. For nongenotoxic carcinogens, in contrast, much more information is needed on issues such as species specificity, threshold effects, and mechanisms of action before rational, safe levels for humans can be set using appropriate safety margins. This has implications both for the design of studies to assess the hazards of chemicals and for the decisions made based on the results of these studies.

At the workshop communication was cultivated between scientists in different disciplines, helping to crystallise scientific opinion on this issue – which is critical to the introduction of new and useful chemicals without jeopardising human safety. We hope that the publication of this book will continue to stimulate discussion and debate on this crucial topic.

Andrew Cockburn
Lewis Sinith
Elaine Aggintin

Table of Contents

List of Contributors

P. Amstad
Department of Carcinogenesis, Swiss Institute for Experimental Cancer
Research, Ch. Boveresses 155, 1066 Epalinges/Lausanne, Switzerland

A. Balmain
CRC Beatson Laboratories, Beatson Institute for Cancer Research,
Garscube Estate, Switchback Road, Bearsden, Glasgow G61 IBD, UK

C. L. Berry
Department of Morbid Anatomy, The Royal London Hospital,
London, E1 1BB, UK

R. Bremner
The Hospital for Sick Children, 555 University Avenue, Toronto Ontario,
M5G 1X8, Canada

K. Brown
CRC Beatson Laboratories, Beatson Institute for Cancer Research,
Garscube Estate, Switchback Road, Bearsden, Glasgow G61 IBD, UK

S. Bryson
CRC Beatson Laboratories, Beatson Institute for Cancer Research,
Garscube Estate, Switchback Road, Bearsden, Glasgow G61 1BD, UK

P. A. Burns
Jack Birch Unit for Environmental Carcinogenesis, Department of Biology,
University of York, Heslington, York YOI 5DD, UK

W. Bursch
Institut für Tumorbiologie-Krebsforschung, Borschkegasse 8a, 1090 Wien,
Austria

P. Cerutti
Department of Carcinogenesis, Swiss Institute for Experimental Cancer
Research, Ch. Boveresses 155, 1066 Epalinges/Lausanne, Switzerland

M. Clarke
CRC Beatson Laboratories, Beatson Institute for Cancer Research,
Garscube Estate, Switchback Road, Bearsden, Glasgow G61 IBD, UK

A. Columbano
Istituto di Patologia Sperimentale, School of Medicine, University of Cagliari,
Cagliari, Italy

N. R. Drinkwater
McArdle Laboratory for Cancer Research, University of Wisconsin Medical
School, 1400 University Avenue, Madison, WI 53706, USA

N. English
Imperial Cancer Research Fund, Molecular Pharmacology Unit, Biomedical
Research Centre, Ninewells Hospital & Medical School, Dundee DDI 9SY,
Scotland, UK

R. Ghosh
Department of Carcinogenesis, Swiss Institute for Experimental Cancer
Research, Ch. Boveresses 155, 1066 Epalinges/Lausanne, Switzerland

B. Grasl-Kraupp
Institut für Tumorbiologie-Krebsforschung, Borschkegasse 8a, 1090 Wien,
Austria

B. Halliwell
Pharmacology Group, King's College, Manresa Road, London SW3 6LX, UK

W. Huber
Institut für Tumorbiologie-Krebsforschung, Borschkegasse 8a, 1090 Wien,
Austria

V. Hughes
Imperial Cancer Research Fund, Molecular Pharmacology Unit, Biomedical
Research Centre, Ninewells Hospital & Medical School, Dundee DDI 9SY,
Scotland, UK

R. L. Jirtle
Department of Radiation Oncology, Duke University Medical Center,
Durham, NC 27710, UK

C. J. Kemp
CRC Beatson Laboratories, Beatson Institute for Cancer Research,
Garscube Estate, Switchback Road, Bearsden, Glasgow G61 IBD, UK

B. G. Lake
BIBRA Toxicology International, Woodmansterne Road, Carshalton,
Surrey SM5 4DS, UK

G. M. Ledda-Columbano
Istituto di Patologia Sperimentale, School of Medicine, University of Cagliari,
Cagliari, Italy

F. Oberhammer
Institut für Tumorbiologie-Krebsforschung, Borschkegasse 8a, 1090 Wien,
Austria

Y. Oya
Department of Carcinogenesis, Swiss Institute for Experimental Cancer
Research, Ch. Boveresses 155, 1066 Epalinges/Lausanne, Switzerland

W. Parzefall
Institut für Tumorbiologie-Krebsforschung, Borschkegasse 8a, 1090 Wien,
Austria

R. Schulte-Hermann
Institut für Tumorbiologie-Krebsforschung, Borschkegasse 8a, 1090 Wien,
Austria

G. Shah
Department of Carcinogenesis, Swiss Institute for Experimental Cancer
Research, Ch. Boveresses 155, 1066 Epalinges/Lausanne, Switzerland

J. A. Swenberg
Departments of Environmental Sciences and Engineering, Pathology
and Curriculum in Toxicology, The University of North Carolina at Chapel
Hill, Chapel Hill, NC 27599, USA

R. W. Tennant
Chief, Laboratory of Environmental Carcinogenesis and Mutagenesis,
National Institute of Environmental Health Sciences, P.O. Box 12233,
Research Triangle Park, NC 27709, USA

G. Thomas
Department of Histopathology, University of Cambridge, Addenbrooke's
Hospital, Hills Road, Cambridge CB2 2QQ, UK

S. Williamson
CRC Beatson Laboratories, Beatson Institute for Cancer Research,
Garscube Estate, Switchback Road, Bearsden, Glasgow G61 IBD, UK

C. R. Wolf
Imperial Cancer Research Fund, Molecular Pharmacology Unit, Biomedical
Research Centre, Ninewells Hospital & Medical School, Dundee DDI 9SY,
Scotland, UK

1 Nongenotoxic Chemical Carcinogens: Evidence for Multiple Mechanisms

R. W. Tennant

1.1 Introduction

In many ways it is easier to generally define nongenotoxic carcinogens by the properties that they lack rather than by those properties they possess. For the purpose of this discussion a minimalist definition of a genotoxic carcinogen will be used. This definition includes chemicals which possess one of the structural alerts, generally indicative of electrophilic potential (Miller and Miller 1977) defined by Ashby (1985; Tennant and Ashby 1991), and which are mutagenic in the *Salmonella* assay (Ames et al. 1973). This definition may exclude some chemicals which may have indirect mutagenic potential, such as those which may undergo specific metabolism in mammalian cells to create mutagenic intermediates, or chemicals that are mutagenic by indirect mechanisms, such as by generating oxidative products which can damage DNA. This paper will focus on the minimalist definition of mutagenicity and how it may aid in identifying and understanding chemicals which operate through nonmutagenic mechanisms to induce cancer.

It is paradoxical that, at a time when the application of molecular genetic methodologies is resulting in the identification of a progressively greater number of genes that are mutated in tumor cells, the rodent bioassay has identified a large number of chemicals which appear to induce cancers by mechanisms which do not involve direct mutagenesis. This paper will discuss how such chemicals have been identified, the types of cancers with which they are associated, and the chemical properties that may be indicative of the mechanisms by which they induce cancer. Finally, I will propose that at least a portion of such nonmutagenic carcinogens act through what may be termed "adaptive" carcinogenic processes and will provide some hypotheses about the mechanisms involved.

1.2 Identification of Nonmutagenic (Nongenotoxic) Carcinogens

All of the chemicals identified and discussed in this presentation derive from the results of rodent bioassays conducted by the U.S. National Toxicology Program (NTP). Additional data from compilations such as those of Gold et al. (1991) or Nesnow et al. (1987) would provide a larger data set for evaluation; however, the use of data derived solely from the NTP rodent bioassay permits a more objective evaluation of results (Huff et al. 1991).

The NTP database is valuable primarily because of the standardized protocol which has been used for over two decades on over 400 chemicals or substances to determine carcinogenic potential. The protocol consists of a subchronic 90-day (13-week) toxicology study which is used to establish doses (generally identified as the maximum tolerated dose, MTD) and which also identifies target tissues based upon nonneoplastic toxic effects. Subsequently, the 2-year (104-week) studies have generally been conducted in Fischer 344 rats and B6C3F1 mice. Fifty animals of each sex/species are exposed to incremental doses (.5 and .25 MTD) of the chemical, and the major differences between each assay relate to the specific chemical or substance and to whether the study is conducted via exposure in food, water, gavage, skin paint, or inhalation. In addition, the bioassay is composed of 50 animals of each sex/species as concurrent controls which are given vehicle by the same route of exposure.

The second important component of the NTP bioassay is the extensive evaluation of each study, which involves complete postmortem examination, both gross and histopathological, including all of the concurrent controls. The program utilizes a standardized nomenclature and peer review of histological material by pathologists to validate the findings of each individual bioassay. This protocol has been generally adhered to, despite the multifold complexities associated with long-term studies in rodents involving exposures at near toxic doses of a variety of chemicals which possess varying physical and biological properties. Abstracts from all of the long-term carcinogenesis studies conducted by the NTP have recently been published (Selkirk and Soward 1993).

The majority of chemicals that will be discussed in this chapter have been described and evaluated (Ashby and Tennant 1991), and the details regarding the specific bioassay of each chemical or substance can be found in the National Toxicology Technical Report Series available through the National Technical Information Service of the U.S. Department of Commerce. These reports are an invaluable source of information on the specific toxic effects, nonneoplastic and neoplastic pathology, and survival and outcome of both subchronic and long-term exposures to chemicals. In addition, this report includes results derived from the in vitro mutagenesis and in vivo cytogenetic studies also conducted by the NTP. Chemicals evaluated in these systems were tested under code and were generally derived from the same sources of chemicals that were used in long-term bioassays. These assays were conducted according to standardized protocols as well and the results evaluated according to predetermined criteria (Tennant and Zeiger 1993).

In previous studies, we have shown that there is a high degree of concordance between specific chemical substructures, termed structural alerts, and the results of *Salmonella* mutagenicity assays (Ashby and Tennant 1991). While approximately 20 discreet structural alerts have been identified, it has been possible to group the majority into three broad chemical groupings or classes. The majority of the structurally alerted chemicals are of the aromatic amino/nitro type compounds which possess NH_2-, $NO-$, $NH-$, $N=0$, $N=N$, etc. joined to an aromatic ring system. The second group consists of natural electrophiles, including chemicals possessing halogen atoms that are in reactive sites on the molecule. These chemicals can generally be considered to be naturally

Table 1. Distribution of *Salmonella* mutagenicity assay results as a function of chemical class and bioassay outcome

Chemical group	Level of effect									
	TC		TS		SM		SS		NEG	
	Sal +	Sal −	Sal +	Sal −	Sal +	Sal −	Sal +	Sal −	Sal +	Sal −
Aromatic amino/nitro-type chemicals	18	0	13	0	5	3	17	1	17	8
Natural electrophiles incl. reactive halogens	11	0	4	2	3	0	6	3	9	7
Minor groups of structurally alerting chemicals	6	1	0	1	0	1	3	4	6	1
Inert halogens	0	4	0	4	0	2	0	14	2	21
Minor structural concerns	1	1	0	5	0	0	1	8	1	18
No structural alerts	0	3	0	2	0	5	0	5	1	44

Sal, *Salmonella;* TC, trans-species, common site; TS, trans-species, single site; SM, single-species, multiple site; SS, single-species, single site.

electrophilic and not requiring metabolic conversion. The third grouping includes a structurally diverse group of chemicals which are classed as nonalerting in structure but with some possible mutagenic potential (e.g., benzaldehyde, benzofuran, furfural, etc.). The actual data supporting mutagenic potential are derived from the results of assays in the *Salmonella typhimurium* strains TA98, 100, etc. A positive response induced in one or more of the strains, with or without exogenous metabolic activation supplement (S9), is sufficient for the chemical to be identified as a mutagen (Tennant et al. 1987).

The high degree of concordance between structural alerts and *Salmonella*-positive chemicals is shown in Table 1. An important aspect of these results is that the concordance between structural alerts and *Salmonella* response could not be improved upon by the substitution or addition of data derived from other in vitro assay systems, such as in vitro cytogenic effects or sister chromatid exchange (SCE) in Chinese Hamster (CHO) cells, etc. We believe that these results mean that, while the *Salmonella* assay and other assays are responsive to chemicals with electrophilic potential, the other genotoxicity assays are also responsive to some other chemical properties that appear to be unrelated to electro-philicity (Tennant and Zeiger 1993). However, there is always the possibility that other assays may be sensitive to specific electrophilic chemicals that do not mutagenize *Salmonella*. This apparent advantage, however, is often offset by the numbers of nonelectrophilic chemicals to which these systems also respond. Therefore, we have focused on the two most concordant properties in order to categorically define mut-agenic chemicals.

Because of the known deficiencies of the S9 metabolic activation systems, there have been extensive efforts to attempt to utilize in vivo genotoxicity assays to help in the identification of true mutagens that may be false negatives in the *Salmonella* assay. Shelby et al. (1993) have recently published the results of an extensive analysis utilizing mouse bone marrow micronucleus assay, and Jackson et al. (1993) have recently published a compilation of data derived from the literature which also indicates that some chemicals may be capable of inducing chromosomal mutations or aneuploidy that can only be detected utiliz-ing such in vivo methods. Therefore, these results show that while the definition of a mutagen may be categorical, some specific mutagens may be excluded. It is unlikely in the foreseeable future that any gener-alized in vivo system can be utilized to identify such specific chemicals, and it is most likely that they will be identified and best understood by detailed studies related to their mode or mechanism of action. I will return to the following point later in this discussion, but it is important to emphasize here that, in defining a chemical as a mutagen, I am not necessarily implying the specific mechanisms by which carcinogenicity may be induced by the chemical. Rather, by this definition I define specific chemical properties which then may be related in some ways to the biological effects of those chemicals.

Table 2 presents a listing of chemicals derived from the NTP bioas-say which are defined as nonmutagenic carcinogens based upon the absence of both structural alerts and *Salmonella* mutagenicity and, where known, on the absence of in vivo cytogenetic effects. This con-stitutes a very structurally and biologically diverse group of agents which, however, does not include some other well-known nonmutagens (e.g., diethylstilbesterol) which have not been studied by the NTP.

Table 2. Nonmutagenic carcinogens

Chemical	Rat		Mouse	
	Male	Female	Male	Female
Aldrin			L	
Allyl isovalerate	HS			LS
11-Aminoundecanoic acid	L UB			
Benzaldehyde			S	S
Benzofuran		K	L LU S	L LU S
Benzyl acetate			S L	S L
o-Benzyl-p-chlorophenol		K		
C.I. Vat yellow 4		HS		
Chlordane (technical grade)			L	L
Chlorendic acid	P L	L	L	
Chlorinated paraffins: Cl2, 60% Cl	HS K L P	HS L TG	L	L TG
Chlorinated paraffins: C23, 43% Cl			HS	
Chlorobenzilate			L	L
Chlorothalonil	K	K		
Cinnamyl anthranilate	K P		L	L
Decabromodiphenyl oxide	L	L		
Di(2-ethylhexyl) phthalate	L	L	L	L
Di(2-ethylhexyl)adipate			L	L
1,4-Dichlorobenzene	K		L	L
p,p'-Dichlorodiphenyldichloroethylene			L	L
Dicofol			L	
N,N'-Diethylthiourea	TG	TG		
3,4-Dihydrocoumarin	K			L
1,4-Dioxane	N	L N	L	L
Diphenylhydantoin				L
Furfural	BD		L	L
Furosemide				MG
Heptachlor			L	L
Hexachloroethane	K		L	L
Hydroquinone	K	HS		L
Isophorone	K			
d-Limonene	K			
Malonaldehyde, sodium salt	TG	TG		
Mercaptobenzothiazole	HS P AG PG	AG PTG		
Mercuric chloride	S			
a-Methylbenzyl alcohol	K			
Methylphenidate HCl			L	L
Nalidixic acid	PG	CG		
Nitrilotriacetic acid (NTA)	K UB UT	K UB UT	K	K
N-Nitrosodiphenylamine	UB	UB		
Pentachloroethane			L	L

Table 2. (cont.)

Chemical	Rat		Mouse	
	Male	Female	Male	Female
Pentachlorophenol	L AG	L CS	L AG	L AG CS
Phenylbutazone		K	L	
Piperonyl sulfoxide			L	
Polybrominated biphenyl mixture	L C	L C	L	L
Probenecid				L
2,3,7,8-Tetrachlorodibenzo-*p*-dioxin	TG	L TG	L	IS L TG
1,1,2,2-Tetrachloroethane			L	L
Tetrachloroethylene	HS	HS	L	L
Triamterene			L	L
1,1,2-Trichloroethane			AG L	AG L
Trichloroethylene			L	L
2,4,6-Trichlorophenol	HS		L	L
Trimethylthiourea		TG		
Tris(2-ethylhexyl)phosphate				L
Zearalenone			PTG	L PTG

Based on 338 NTP rodent bioassays defined es nonmutagens based upon structural alert (SA) negative, *Salmonella* mutagenesis (SAL) negative, and where data were available of a negative response in an in vivo cytogeneties assay or for induction of micronuclei.

AG, adrenal gland; BD, bile duct; C, cholangioma; CG, clitoral gland; CS, circulatory system; HS, hematopoietic system; IS, integumentary system; K, kidney; L, liver; LU, lung; MG, mammary gland; N, nose; P, pancreas; PG, preputial gland; PTG, pituitary gland; S, stomach; TG, thyroid gland; UB, urinary bladder; UT, uterus.

1.3 Patterns of Tumors Induced by Nonmutagens

In a previous study, the patterns of tumors induced in bioassays were analyzed in relation to the species, sex, and site distribution of tumors for chemicals classified as showing either *clear* or *some* evidence of carcinogenicity (Ashby and Tennant 1991). Subsequently, the often remarkable differences in the relative distribution of tumors induced by different chemicals provided a conceptual basis for stratification of rodent carcinogenicity bioassay results (Tennant 1993). Table 2 shows the distribution of tumors for nonmutagens and suggests two important general properties of nonmutagens: First, they tend to produce a higher proportion of cancers limited to a single species or to a single site than

Table 3. Number of sites of neoplasia for SS category chemicals ($n = 159$)

Sites	Rat	Mouse
Kidney	8	–
Urinary bladder	7	–
Thyroid gland	4	1
Hematopoietic system	3	2
Liver	3	21
Intestine/colon	1	1
Multiple organ sites	1	–
Pancreas	1	–
Stomach	1	1
Spleen	1	–
Uterus	1	1
Circulatory system	–	2
Mammary gland	–	1
Nose	–	1

SS, single-species, single site.

mutagenic carcinogens; second, for the respective species, the highest proportion of the tissue-specific effects of these carcinogens involves relatively few sites. The stratification scheme that has been proposed emphasizes the ability of chemicals to induce carcinogenic effects across sexes and species and to induce tumors in at least one common site in both rats and mice. Chemicals that demonstrated this pattern of tumorigenesis are classified as transspecies carcinogens (TC). The second category of activity is chemicals that induce transspecies effects in at least one sex of both rats and mice, but where no common site is involved (TS). The third category identifies chemicals that induce carcinogenesis in a single species, but at multiple sites (SM), and the fourth category represents those chemicals that induce tumors in a specific site in only one sex of one of the species (SS). The relative distribution of mutagens for the TC and SS categories are 78% and 40%, respectively. That is, 78% of the chemicals which demonstrated transspecies carcinogenic potential were mutagens, whereas 60% of the chemicals that were carcinogenic only at a single site in a single sex/species were nonmutagens. The purpose of proposing this stratification was to provide a means of judging the relative biological potency of carcinogens, since the current NTP classification scheme distinguishes only between

Table 4. Stratification of nonmutagenic carcinogens

Category	Total ($n = 56$)	
	(n)	$(\%)$
Trans-species, common sites (TC)	7	12.5
Trans-species, single site (TS)	12	21.4
Single-species, multiple sites (SM)	6	10.7
Single-species, single site (SS)	31	55.4

"clear" and "some" evidence of carcinogenesis. A chemical that is identified in either of these categories is regarded as a carcinogen. It is also interesting that the average MTD for chemicals falling in the SS category is approximately twice that of chemicals which are classified as TC. This means that, on average, chemicals that have the capacity to induce transspecies carcinogenesis tend to be much more toxic. An important implication of this stratification scheme is that chemicals that have the capacity to induce transspecies carcinogenesis are more likely to be carcinogenic in other species than are the chemicals that show the highly restricted patterns of carcinogenic potential (Tennant 1993). These results may also have important implications for our understanding of the mechanisms by which nonmutagenic carcinogens may function. It is highly probable that the site specificity of the carcinogenic nonmutagens are heavily influenced by the genotype of mice and rats used in the bioassay. For example, examination of the distribution of site-specific effects in mice reveals that the mouse liver is a predominant site of carcinogenic effect (Table 3). The identification of genetic loci (*hcs*) (Drinkwater and Ginsler 1986) which influence chemical carcinogenesis in the mouse liver and that are contributed by the C3H parent in this F1 cross suggests that the *hcs* locus may be an important determinant of carcinogenesis in the mouse.

The distribution of the nonmutagenic carcinogens as a function of stratification category is shown in Table 4. Two thirds of the carcinogens induce single species effects, with over half inducing tumors at single sites in a single species.

The Fischer F344 rat has been less well characterized genetically and no specific loci have been identified which influence the patterns of spontaneous tumorigenesis which are well documented to occur in

Table 5. Toxicity/carcinogenicity relationships for some nonmutagens which induced renal tumors

Chemical	Chronic toxicity Rat	Mouse	Carcinogenicity Rat	Mouse
Benzofuran	K, P, S	L, LU, S	K female	L, LU, S
Ochratoxin A	K	nd	K male female, MG	nd
Hexachloroethane[a]	K	nd	K male	nd
d-Limonene[a]	K	L	K male	–
Hydroquinone	K	L, TG	K male, HS	L
Phenylbutazone	K, S, AG	L	K female	L male
Furosemide	K	K	–	MG
α-Methylbenzyl alcohol	K	–	K male	–

See Table 2 for site legend.
[a]Associated with α2μ-globulin.

Fischer rats. While the kidney and urinary bladder are sites of relatively low spontaneous tumor incidence (Haseman et al. 1985), it is clear from the data presented in Table 2 that these sites are frequent targets for the effects of nonmutagenic carcinogens that are specific to the rat. These results suggest that one or more genetic loci may influence kidney and bladder carcinogenesis in the rat, and studies to identify such loci may prove very fruitful. These results, however, also suggest that the action of nonmutagenic carcinogens is not simply to "promote" the development of spontaneous tumors at these sites in the rat. Rather it may be that some specific chemical properties are influenced by the tissue-specific expression of certain genes, thereby targeting the action of the chemical.

Although spontaneous tumors of the renal epithelium of F344 rats are not common, age-related effects generally characterized as nephropathy are quite common (Boorman et al. 1990). Since it was possible that this condition could influence the frequency of chemical-induced renal neoplasia, we evaluated some recently tested chemicals which induced renal tumors. From the results shown in Table 5 it is clear that chemicals which induce renal tubular cell cancers are also nephrotoxic. An exception, however, was furosemide which showed renal toxicity in both rats and mice, but induced only mammary gland neoplasias in mice and no neoplasias in rats. Of the seven chemicals inducing renal tumors in rats, the induction of α2-μ globulin has been associated only with hexachloroethane and d-limonene (Borghoff et al. 1990). Other chemicals that induced renal tumors only in male rats did not show evidence of $\alpha 2$-μ globulin deposition. These observations, together with the structure and biological diversity of the chemicals, suggest that there are probably multiple mechanistic pathways for the induction of rat renal tumors by nonmutagens (Barrett and Huff 1991).

The induction of liver carcinogenesis in mice also presents evidence of mechanistic complexity. The existence of at least one locus influencing suseptibility has been reported by Drinkwater and Ginsler (1986) and Dragani et al. (1991). The influence of this locus in the B6C3F1 hybrid mice may be at least partially responsible for the high frequency with which the liver is a site of chemical carcinogenesis. This locus is not solely responsible since a relatively high proportion of carcinogens also induce liver tumors in the F344 rat. The F344 rat has a much lower incidence of spontaneous liver tumors and no suseptibility genes have yet been identified.

The role of the *ras* gene in the hepatocarcinogenesis in the B6C3F1 mouse has also been extensively evaluated. Activated Ha ras has been found in spontaneous tumors as well as in some chemical-induced tumors (Reynolds et al. 1987; Fox et al. 1990). Specific mutations in the 61st codon have been identified in a high proportion (greater than 60%) of spontaneous tumors, and most carcinogens were found to increase mutation frequency. Tumors induced by some nonmutagens showed a mutation frequency similar to those in spontaneous tumors. Therefore, both spontaneous and nonmutagen induced tumors may develop in the absence of *ras* activation. The origin of the spontaneous mutations is not understood but is probably related to the genotype of the C3H parent which may also be partially responsible for the high spontaneous incidence of hepatocarcinogenesis. No consistent pattern was observed for *ras* mutations of tumors induced by some nonmutagenic chemicals (phenobarbital, chlorfrom, and cyprofibrate). The mutations identified in those tumors indicated that the *ras* mutations were of spontaneous origin and that the majority of tumors did not involve a mutated *ras* gene.

The problem of mouse liver tumors is thus complex and may involve other genetic elements that have not yet been recognized. Based upon an evaluation of over 300 chemicals (Ashby and Tennant 1991), there were no obvious structural features of chemicals that were associated with

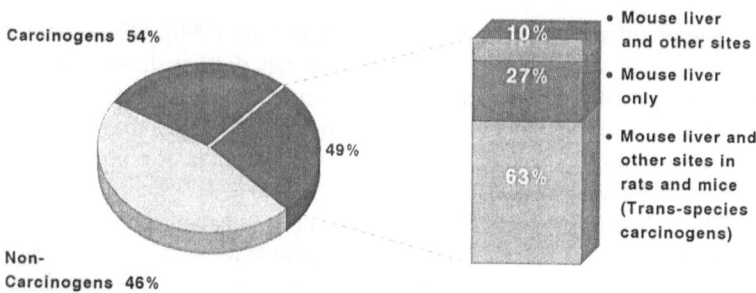

Fig. 1. Distribution of mouse liver carcinogens

Table 6. Proportion of *Salmonella* mutagens among stratified mouse hepatocarcinogens

	Number (%)
Mouse liver only	8/21 (38)
Mouse liver and other sites in mice	3/8 (38)
Mouse liver and other sites in rats and mice	31/50 (62)

hepatocarcinogenesis or hepatotoxicity. An evaluation of the distribution of chemicals that induce mouse tumors as a function of stratified responses is presented in Fig. 1. It demonstrates that the highest proportion of mouse liver carcinogens are in fact transspecies carcinogens that induce tumors at other sites in rats and mice. Table 6 shows the proportion of mutagens among the three categories of hepatocarcinogens and predictably shows that approximately 2/3 of the transspecies carcinogens that induce mouse liver tumors are mutagens and that the highest proportion of nonmutagens is represented among the chemicals that induce tumors only in mice. Therefore, these chemical and biological characteristics of mouse liver carcinogens strongly suggests the existence of multiple-mechanistic pathways.

1.4 Mechanisms of Nonmutagenic Carcinogenesis

The biological and chemical complexity of the nonmutagenic carcinogens identified in this study argue strongly against any unified theory of nongenotoxic carcinogenesis (Tennant et al. 1990). Rather, all of the evidence suggests that the concept of nongenotoxic carcinogenesis is a default definition when we are characterizing chemicals that lack certain common structural or biological features. It is important to recognize that virtually all of the nongenotoxic carcinogens and a very high proportion of the genotoxic carcinogens identified in bioassays are recognized only at the end of protracted periods of exposure (2 years). Thus far, no consistent pattern relating to either induced cellular proliferation, indirect oxidative damage, or tumor promotion can be ascribed to a majority of these chemicals (Tennant and Zeiger 1993). While it has been proposed that such late occurring tumors simply reflect the relatively low potency of nonmutagenic carcinogens, the results might also reflect an interplay of adaptive cellular or molecular mechanisms to the consequences of protracted chemical exposure. Adaptive mechanisms to acute toxicity have been defined for many chemicals, where altered patterns of transcription and gene expression are utilized to allow cells to survive in the presence of toxic exposure. A dramatic example of the adaptive capacity of animal cells is found in a process of cell culture whereby cells taken from the regulated environment of the whole animal are required to adapt and survive on plastic surfaces in the presence of artificial medium components, in serum derived from foreign species, and in the absence of intracellular communication and other regulatory signals. These adaptive processes, in many ways, resemble the neoplastic process (Farber and Rubin 1991). Cells may downregulate the expression of a variety of differentiated functions, lose contact inhibition, acquire extended lifespan, alter or lose the capacity for terminal differentiation, and frequently undergo sustained unregulated growth. An important aspect of this process is that it is a series of stochastic events in which progeny of the adaptive cells exhibit the same characteristics. The consequence of such adaptation is often the ultimate loss of genomic stability and the acquisition of aneuploidy. It is important to note that the capacity for adaptation exists within virtually all mammalian cells that are not terminally differentiated and has its basis in changes in the pattern of gene transcription. Where such changes result in heritable phenotypic change, the probability of neoplasia would significantly increase. The direct implication of this hypothesis is that some types of cancer, particularly those that arise from exposure to nonmutagens, are a consequence, at least in part, of altered transcription rather than of mutation. A major question is whether dysregulation, overexpression, amplification or inappropriate expression of specific growth or differentiation-related genes can mimic the consequences of induced mutations in the same genes and whether those changes can be transmitted heritably. The consequences of sustained exposure to chemicals which could function as pseudoligands for receptors involved in important growth- and differentiation-related regulatory pathways would provide a plausible mechanism by which such heritable phenotypic changes could be sustained (Green 1993). Since genomic instability appears to be a consistent consequence of development of the malignant phenotype, the search for induced adaptive changes will require studies conducted during early phases of induction of cancers by nonmutagens.

References

Ames BN, Durston WE, Yamasaki E, Lee FD (1973) Carcinogens are mutagens: a simple test system for combining liver homogenates for activation and bacteria for detection. Proc Natl Acad Sci USA 70:2281–2285

Ashby JA (1985) Fundamental structural alerts to potential carcinogenicity and non-carcinogenicity. Environ Mutagen 7:919–921

Ashby J, Tennant RW (1991) Definitive relationships among chemical structure carcinogenicity and mutagenicity for 301 chemicals tested by the U.S. NTP. Mutat Res 257:229–306

Barrett JC, Huff J (1991) Cellular and molecular mechanisms of chemically induced renal carcinogenesis. Renal Fail 13:211–225

Boorman GA, Eustis SL, Elwell MR, Montgomery CA, Mackenzie WF (1990) Pathology of the Fischer rat. Academic, New York, pp 127–152

Borghoff S, Short BG, Swenberg JA (1990) Biochemical mechanisms and pathobiology of $\alpha 2\mu$-globulin nephropathy. Annu Rev Pharmacol Toxicol 30:349–367

Dragani TA, Manenti G, Della Porta G (1991) Quantative analysis of genetic susceptibility to liver and lung carcinogenesis in mice. Cancer Res 51:6292–6303

Drinkwater NR, Ginsler J (1986) Genetic control of hepatocarcinogenesis in C567Bl/6N and C3H/hen inbred mice. Carcinogenesis 7:1701–1707

Farber E, Rubin H (1991) Cellular adaptation in the origin and development of cancer. Cancer Res 51:2751–2761

Fox TR, Schumann AM, Watanabe PG, Yano BL, Maher VM, McCormick JJ (1990) Mutational analysis of the H-ras oncogene in spontaneous C57Bl/6C3H/He mouse liver tumors and tumors induced with genotoxic and non-genotoxic hepatocarcinogenesis. Cancer Res 50:4014–4019

Gold LS, Slone TH, Manley NB, Bernstein L (1991) Target organs in chronic bioassays of 533 chemical carcinogens. Environ Health Perspect 93:233–246

Green S (1993) Nuclear receptors and chemical carcinogenesis. Trends Pharmacol Sci 13:251–255

Haseman JK, Huff JH, Rao GN, Arnold JE, Boorman GA, McConnell EE (1985) Neoplasms observed in untreated and corn oil gavage control groups of F344/N rats and (C57Bl/6NC3H/Hen) F1 (B6C3F1) mice. J Natl Cancer Inst 75:975–984

Huff J, Haseman J, Rall D (1991) Scientific concepts, value and significance of chemical carcinogenesis studies. Annu Rev Pharmacol Toxicol 31:621–652

Jackson MA, Stack H, Waters MD (1993) The genetic toxicology of putative non-genotoxic carcinogens. Mutat Res 296:241–277

Miller JA, Miller EC (1977) Ultimate carcinogens as reactive mutagenic electrophiles. In: Hiatt HH, Watson JD, Weinstein BA (eds) Origins of human cancer. Cold Spring Harbor Laboratoy Press, Cold Spring Harbor, New York, pp 602–628

Nesnow S, Argus M, Bergman H, Chu K, Frith C, Helms T, McGaugny R, Ray V, Slaga TN, Tennant RW, Weisburger E (1987) Chemical carcinogens: a review and analysis of the literature of selected chemicals and the establishment of the gene-tox carcinogen database – a report of the U.S. environmental protection agency gene-tox program. Mutat Res 185:1–195

Reynolds SH, Stowers SJ, Patterson RM, Maronpot RR, Aaronson SA, Anderson MW (1987) Activated oncogenes in B6C3F1 mouse liver tumors: implication for risk assessment. Science 237:1309–1316

Selkirk JK, Soward SM (1993) Compendium of abstracts from long-term cancer studies reported by the National Toxicology Program. Environ Health Perspect 101:1–291

Shelby MD, Erexson GL, Hook GJ, Tice RR (1993) Evaluation of a three-exposure mouse bone marrow micronucleus protocol: results with 49 chemicals. Environ Mol Mutagen 21:160–179

Tennant RW (1993) Stratification of rodent carcinogenicity bioassay results to reflect relative human hazard. Mutat Res 286:111–118

Tennant RW, Ashby J (1991) Classification according to clinical structure, mutagenicity to Salmonella and level of carcinogenicity of a further 39 chemicals tested for carcinogenicity by the U.S. National Toxicology Program. Mutat Res 257:209–227

Tennant RW, Zeiger E (1993) Genetic toxicology: current status of methods of carcinogen identification. Environ Health Perspect 100:307–315

Tennant RW, Margolin BH, Shelby MD, Zeiger E, Haseman JK, Spalding J, Caspary W, Resnick M, Stasiewicz S, Anderson B, Minor R (1987) Prediction of chemical carcinogenicity in rodents from in vitro genetic toxicity assays. Science 236:933–941

Tennant RW, Spalding J, Stasiwicz S, Ashby J (1990) Prediction of rodent carcinogenicity for 44 chemicals currently being evaluated by the U.S. National Toxicology Program. Mutagenesis 5:3–14

2 Oxidative Damage and Carcinogenesis

P. Amstad, R. Ghosh, G. Shah, Y. Oya, and P. Cerutti

2.1 Introduction

Substantial evidence has suggested that free radicals, particularly oxygen radicals, play an important role in several stages of carcinogenesis. Oxidants are ubiquitous in our natural environment but they can also be formed in the tissue by endogenous cellular mechanisms (Cerutti 1985; Kozumbo et al. 1985; Cerutti and Trump 1991). Oxidants can introduce structural damage to DNA, leading to chromosomal aberrations and point mutations. Point mutations in cancer-related genes such as the ras-family protooncogenes (Bos 1989) and the p53 tumor suppressor gene (Hollstein et al. 1991; de Fromentel and Sossi 1992) represent the most frequent genetic changes in human malignancies and at least part of them may be caused by oxidants. Besides these "genotoxic" effects, oxidants activate signal transduction pathways which lead to the modulation of the expression of growth- and differentiation-related genes (Crawford et al. 1988; Shibanuma et al. 1988; Cerutti et al. 1990). However, unlike growth factors oxidants always induce macromolecular damage, cytotoxicity, and cell killing. The effect of oxidants are influenced by the cellular antioxidant defenses (Cerutti et al. 1988; Amstad and Cerutti 1990) which consist of low-molecular-weight antioxidants and several antioxidant enzymes. The biological consequences of the exposure to an oxidant carcinogen may vary with the dose, the type of oxidant, and the tissue because it is the result of the superposition of effects on multiple cellular targets.

Chronic tissue injury by physical and chemical irritants frequently results in inflammation which is accompanied by the infiltration of phagocytic leukocytes (Creasia and Nettesheim 1974; Mass et al. 1985; Alexander-Williams 1976; Argyris and Slaga 1981). Phagocytic leukocytes produce a highly complex mixture of growth and differentiation factors as well as biologically active arachidonic acid metabolites (Edginton et al. 1987). In addition, they have the capacity to release large amounts of active oxygen (AO) in an oxidative burst (Badwey and Karnovsky 1990). Current evidence suggests that AO and arachidonic acid metabolites are important in tumorigenesis. Low-molecular-weight antioxidants, antioxidant enzymes, and anti-inflammatory agents that inhibit arachidonic acid metabolism are anticarcinogenic in several experimental systems (Slaga et al. 1983; Viaje et al. 1977; Kensler et al. 1983). The hypothesis that AO from phagocytes may be an important carcinogen is supported by the finding that an extracellular burst of AO produced by xanthine/xanthine oxidase (X/XO) is a potent promoter for initiated mouse embryo 10T1/2 fibroblasts and mouse epidermal JB6 cells (Zimmerman and Cerutti 1984; Muehlematter et al. 1988). While H_2O_2 itself is a weak promoter for initiated mouse skin, several xenobiotic organic endo- and hydroperoxides possess considerable potential as promoters and progressors (Slaga et al. 1983; Gindhart et al. 1985; O'Connell et al. 1986).

Polycyclic aromatic hydrocarbons (PAH) represent important etiological agents in lung cancer induced by tobacco smoke. They are metabolically activated to epoxy intermediates which form covalent adducts to DNA. In addition PAH metabolites and other aromatic compounds which have the potential to form quinoid intermediates can induce oxidative DNA damage.

As mentioned above, the cellular antioxidant defenses are bound to play a role in oxidant carcinogenesis. Epidemiological studies on serum antioxidants and diet suggest that elevated levels of vitamin E and beta-carotene reduce mortality due to cancer in the lung and colon (Menkes et al. 1986; Gey 1987). In view of the multiple stages and targets where oxidants can act in carcinogenesis, it may not be astonishing that the effect of the antioxidant defense depends on the cell type and tissue. For example, high antioxidant capacity is expected to protect the DNA from oxidative damage and mutagenesis but at the same time it may protect "initiated" cells from excessive oxidant toxicity and favor their clonal expansion in tumor promotion (Amstad and Cerutti 1990; Cerutti 1988; Crawford et al. 1989).

In this chapter we review progress made in our laboratory in the understanding of the role of the antioxidant enzymes superoxide dismutase (SOD), catalase and glutathione peroxidase (GPx) in oxidant-induced genotoxicity. In addition we report on our work on ultraviolet B light (UVB)-induced carcinogenic effects mediated in part by oxidative stress.

2.2 The Effect of the Antioxidant Defense on Oxidant-Induced Chromosome and DNA Breakage

Oxidants can cause permanent structural damage to DNA as well as transient changes in gene expression. The sensitivity of the genome to oxidants is modulated by the cellular antioxidant defense. In order to study the effect of the major antioxidant enzymes on genome vulnerability we have constructed genetic variants of promotable mouse epidermal cells JB6 clone 41 by transfection with complementary DNA (cDNA) coding for human Cu,Zn-SOD, catalase and bovine Se-GPx. Different resistance cassettes were used for the construction of each of the three expression vectors, allowing the preparation of single- and double transfectants with increased complements of one or two enzymes. This allows it to dissect the individual contributions of these interacting enzymes to the overall antioxidant defense. The transfectants were completely analyzed on the molecular and biochemical level (Amstad and Cerutti 1990; Amstad et al. 1991; P. Amstad, R. Moret, and P. Cerutti, unpublished).

Below we describe the effects of the modulation of the antioxidant defense on oxidant-induced chromosome- and DNA damage.

2.2.1 Cu,Zn-SOD Transfectants Are Sensitized to Oxidant-Induced DNA Damage While Catalase and GPx Transfectants Are Protected

Cytogenetic analysis revealed that the Cu,Zn-SOD transfectants SOD3 and SOD15 were sensitized to chromosomal damage induced by oxidant. Two- to threefold higher SOD activities resulted in a three- to fivefold increase in total chromosomal aberrations following exposure to xanthine/xanthine oxidase (X/XO) relative to the parent strain (aberrations scored: breaks, isobreaks, gaps, isogaps, exchanges; Y. Oya and P. Cerutti, unpublished). Similarly, the SOD transfectants were sensitized to X/XO-induced DNA strand breakage measured by the alkaline elution method (Amstad et al. 1991).

In contrast, stable transfectants with increased levels of CAT were protected from X/XO-induced chromosome and DNA breakage. For example, in transfectant CAT4 with catalase activity increased approximately threefold, the oxidant-induced total chromosomal aberration frequency was reduced fourfold relative to the parent strain (Y. Oya and P. Cerutti, unpublished). As expected, catalase transfectants were also more resistant to oxidant-induced DNA strand breakage. These results indicate that the balance between SOD and catalase plays a crucial role for the overall vulnerability of the genome to a mixture of $O_2^{\bullet-}$ and H_2O_2 produced extracellularly by X/XO (Amstad et al. 1991).

In a recent study we evaluated the effect of increasing the cellular complement in GPx on the oxidant sensitivity of the parent strain JB6 clone 41 and its SOD-transfectants SOD3 and SOD15. Sensitivity to DNA strand breakage and killing by X/XO was reversely related to the ratio of activities GPx over SOD. A GPx transfectant with a GPx/SOD ratio of 3.8 was very strongly protected. The hypersensitivity of the SOD clones with GPx/SOD ratios of 0.4 was corrected or overcorrected by secondary transfection with bovine seleno-GPx, resulting in increased activity ratios GPx/SOD of 1.0 to 2.4. Our results indicate that the ratio GPx/SOD determines the resistance of cells to oxidant-induced damage to the genome and cell killing (P. Amstad, R. Moret, and P. Cerutti, unpublished). X/XO produces a large burst of active oxygen close to the cell surface and it is conceivable that lipid peroxidation in the membrane and the formation of long-lived clastogenic products are on the pathway to DNA breakage (Cerutti et al. 1983; Ochi and Cerutti 1987; Lewis et al. 1986). Indeed, in part the protective action of GPx might be due to its capacity to destroy clastogenic lipid hydroperoxides.

A satisfactory interpretation of these results requires the understanding of the reasons for the toxicity of increased levels of Cu,Zn-SOD. The compensatory effect of catalase and GPx suggests that overproduction of H_2O_2 by $O_2^{\bullet-}$ dismutation might be responsible for SOD toxicity (Amstad et al. 1991; Mao et al. 1983). Alternatively, overscavenging of hydroperoxy radical $HO_2^{\bullet}$ (the conjugate acid of $O_2^{\bullet-}$) by excess SOD may reduce radical chain termination and result in increased lipid peroxidation (Omar and McCord 1990). Finally, the inherent peroxidatic activity of Cu,Zn-SOD could play a role (Hodgson and Fridovich 1975). However, the fact that both excess Cu,Zn- and Mn-SOD have been shown to be toxic in an ischemia–reperfusion model argues against this possibility since the latter enzyme lacks peroxidatic activity (Omar and McCord 1990). We do not imply that DNA is the immediate target for attack by H_2O_2 (Cantoni et al. 1989) or its radical derivatives. The fact that the chelation of intracellular Ca^{2+} strongly inhibits DNA breakage by H_2O_2 suggests that the activation of Ca^{2+}-dependent endonucleases plays a role. It should be noted that evidence in the literature supports the notion that overexpression of SOD can sensitize rather protect cells from oxidant stress (Scott et al. 1987; Bloch and Ausubel 1986; Elroy-Stein et al. 1986).

2.3 Oxidant Stress Induced by UVB Contributes
to Its Carcinogenic Effect

The UVB portion in the wavelength range from 290–320 nm possesses the highest potency for the induction of skin cancer (Coohill et al. 1987; Epstein 1978; Forbes et al. 1978). Whereas UVC (190–290 nm) preferentially causes damage to DNA via electronic excitation, UVB interacts with multiple cellular targets and appears at least in part to act by oxidative mechanisms. This is supported by the observation that UVB efficiently induces lesions of the 5,6-dihydroxy-dihydrothymine type in DNA as well as single strand breaks (Hariharan and Cerutti 1977; Hirschi et al. 1981; Niggli and Cerutti 1983).

The question arises whether UVB and oxidants have the capacity to introduce mutations into protooncogenes and tumor suppressor genes. For the detection of base pair changes in a minority of cells without the selection of phenotypically altered cells we have developed the restriction fragment length polymorphism/polymerase chain reaction (RPLP/PCR) protocol which measures mutations in restriction endonuclease recognition sequences. This "genotypic" mutation system is being applied to UV- and oxidant-induced mutagenesis of cancer-related genes in human cells (Chiocca et al. 1992; Pourzand and Cerutti 1993).

The overall biological consequences of UVB are expected to result from the superposition of its genetic and epigenetic effects. Besides causing DNA damage, it activates signal transduction pathways which originate at the plasma membrane and which involve kinases and phosphatases known to participate in the mitogenic response to certain growth factors (Forbes et al. 1978; Devary et al. 1992; Radler-Pohl et al. 1993). UVB modulates the expression of several growth-related genes, among them the immediate early genes c-fos and c-jun (Roinai et al. 1988; Kaina et al. 1990; Devary et al. 1991; Shah et al. 1993). The reason for the high carcinogenicity of UVB might be the fact that it induces structural damage to the genome and at the same time stimulates epidermal proliferation. Since UVB induces oxidative stress, it is to be expected that the cellular antioxidant defense modulates its action.

2.3.1 Mechanism of c-fos Induction by UVB

In order to understand the growth-stimulatory effect of UVB we are studying the mechanism of transcriptional induction of immediate early protooncogenes in mouse epidermal JB6 cells and made the following observations. UVB is a moderate transcriptional inducer of c-fos and c-jun. It induces a biphasic response of c-fos with an early peak at 30–60 min and a second, broader peak at 7–8 h. Only the early phase of expression is suppressed by inhibitors of 2'-monophosphoadenosine 5'-diphosphoribose (ADPR)-transferase. We propose that the two phases of c-fos induction by UVB occur by quite different mechanisms. The early phase requires polyADP ribosylation of chromosomal proteins for the resealing of UVB-induced DNA breaks which otherwise suppress transcription. Experiments with conditioned media from UVB-irradiated cells indicate that an autocrine factor may be responsible for the late phase of c-fos induction. These features of c-fos induction are characteristic for UVB and have not been observed for stimulation by serum nor phorbolester.

We conclude that the action of UVB is the superposition of specific mutational changes in cancer-related genes, general genotoxicity and growth factor-like epigenetic effects (Shah et al. 1993).

2.3.2 The Effect of the Antioxidant Defense on DNA Strand Breakage and Transcriptional Induction of c-fos by UVB

In order to evaluate the capacity of the antioxidant defense to protect the genome from UVB-induced damage we compared the sensitivities to DNA strand breakage of parent JB6 cells and its stable catalase, SOD, and GPx transfectants. Catalase transfectants with catalase activities increased three- to fourfold were strongly protected. Similar results were obtained with the bovine GPx transfectant GPx20 which possesses GPx activity which is elevated threefold relative to the parent strain. In contrast, the two Cu,Zn-SOD transfectants SOD3 and SOD15 possessed sensitivities which were comparable to the parent strain. On the basis of enzymatic specificities of the transfected enzymes we conclude that UVB-induced DNA breakage is at least in part mediated by the formation of H_2O_2 and possibly organic hydroperoxides but that $O_2^{\cdot-}$ is not directly involved (P. Amstad, R. Moret and P. Cerutti, unpublished).

Unrepaired DNA breaks may be incompatible with efficient transcription of c-fos. Indeed, the fact that UVB only moderately induces c-fos may be the consequence of its genotoxic effect which is superimposed on its potential to activate the necessary signal transduction pathways. According to this model reduced strand breakage in GPx transfectants would be expected to enhance the transcriptional induction of c-fos. We therefore compared the increase in c-fos message induced by UVB and serum between the parent JB6 clone 41 and GPx20 by northern analysis. In agreement with our model c-fos expression by UVB was enhanced threefold in GPx20 but slightly reduced in the SOD clones SOD3 and SOD15. The antioxidant status had no significant effect on c-fos expression by serum (P. Amstad, R. Moret and P. Cerutti, unpublished).

2.3.3 UVB-Induced DNA Breaks Have a Long-Range Effect on Chromatin Structure Which Inhibits c-fos Transcription

As already mentioned, polyADP ribosylation of chromosomal proteins is required for the efficient resealing of DNA breaks and consequently its inhibition with 3-amino-benzamide (3-AB) suppressed c-fos induction by UVB and oxidants. However, the fos gene and its regulatory sequences represent a very small target and most likely do not contain a significant number of breaks at moderate UVB or oxidant doses (Shah et al. 1993). It is more likely that unrepaired breaks exert a long-range effect on chromatin conformation which is incompatible with efficient transcriptional induction. We have tested this model in experiments comparing the effect of ADPR transferase inhibition on the UVB induction of the endogenous c-fos gene, of a stably integrated p-fos–CAT construct (CAT, chloramphenicol acetyltransferase) containing the full length 5'-regulatory sequences of c-fos and of the same transiently transfected p-fos–CAT construct. In contrast to the stably integrated vector, the transiently transfected extrachromosomal vector does not assume a native, higher-order chromatin structure and is not susceptible to long-range effects by DNA breaks. The preparation of stable transfectants containing the fos regulatory sequences linked to the CAT reporter gene has been described (Forbes et al. 1978). Serum-starved cultures were either irradiated with UVB or treated with serum in the presence or absence of 3-AB and total RNA extracted after 60 min. RNAse protection analysis of CAT-RNA expression indicated that 3-AB suppressed the induction of the CAT-reporter gene by UVB but not serum in the stable p-fos–CAT transfectants. It should be noted again that serum does not cause DNA strand breakage in JB6 cells. In contrast, 3-AB had no effect on the induction of the identical, transiently transfected p-fos–CAT construct. As expected from the previous results mentioned above, 3-AB suppressed the UVB induction of the endogenous c-fos gene in the stably and transiently transfected cultures (Ghosh et al. 1993). These results support our proposition that DNA breaks exert a long-range effect on chromatin conformation which interferes with fos transcription. It is well documented that chromatin undergoes conformational changes in regions of active transcription (Shalder et al. 1980; Weintraub 1987). For the case of the c-fos gene it has been demonstrated that a transient gradient of increased DNaseI sensitivity extends hundreds of base pairs upstream and downstream from the serum responsive element enhancer motif when HeLa cells are stimulated with serum (Feng and Villeponteau 1990, 1992). It is conceivable that this type of conformational change of chromatin cannot be established in the presence of unrepaired DNA breaks.

Acknowledgements. This work was supported by the Swiss National Science Foundation, the Swiss Association of Cigarette Manufacturers, and the Association for International Cancer Research.

References

Alexander-Williams J (1976) Inflammatory disease of the bowel. Dis Colon Rectum 19:579–581

Amstad P, Cerutti P (1990) Genetic modulatin of the cellular antioxidant defense capacity. Environ Health Perspect 88:77–82

Amstad P, Peskin A, Shah G, Mirault ME, Moret R, Zbinden I, Cerutti P (1991) The balance between Cu,Zn-superoxide dismutase and catalase affects the sensitivity of mouse epidermal cells to oxidate stress. Biochemistry 30:9305–9313

Argyris T, Slaga T (1981) Promotion of carcinomas by repeated abrasion in initiated skin of mice. Cancer Res 41:5193–5195

Badwey J, Karnovsky M (1990) Active oxygen species and the functions of phagocytic leucocytes. Annu Rev Biochem 49:695–726

Bloch C, Ausubel F (1986) Paraquat-mediated selection for mutations in the Mn-SOD gene SODA. J Bacteriol 168:795–798

Bos J (1989) ras Oncogene in human cancers. A review. Cancer Res 49:4682–4689

Cantoni O, Sestili P, Cattabeni F, Bellomo G, Pou S, Cohen M, Cerutti P (1989) Calcium chelator Quin 2 prevents hydrogen-peroxide-induced DNA breakage and cytotoxicity. Eur J Biochem 182:209–212

Cerutti P (1985) Prooxidant states and tumor promotion. Science 227:375–381

Cerutti P (1988) Commentary: response modification creates promotability in multistage carcinogenesis. Carcinogenesis 9:519–526

Cerutti P, Trump B (1991) Inflammation and oxidative stress in carcinogenesis. Cancer Cells 3:1–7

Cerutti P, Emerit I, Amstad P (1983) Membrane mediated chromosomal damage. In: Weinstein IB et al. (eds) Genes and proteins in oncogenesis. Academic, London

Cerutti P, Fridovich I, McCord J (eds) (1988) Oxyradicals in molecular biology and pathology. Liss, New York

Cerutti P, Larsson R, Krupitza G (1990) Mechanisms of oxidant carcinogenesis. In: Harris C, Liotta L (eds) Genetic mechanisms in carcinogenesis and tumor progression. Wiley-Liss, New York

Chiocca S, Sandy M, Cerutti P (1992) Genotypic analysis of N-ethyl-N-nitrosourea-induced mutations by Taq1 restriction fragment length polymorphism/polymerase chain reactin in the c-H-ras1 gene. Proc Natl Acad Sci USA 89:5331-5335

Coohill T, Peak M, Peak J (1987) The effects of the ultraviolet wavelengths of radiation present in sunlight on human cells in vitro. Photochem Photobiol 46:1043–1050

Crawford D, Zbinden I, Amstad P, Cerutti P (1988) Oxidant stress induces the protooncogenes c-fos and c-myc in mouse epidermal cells. Oncogene 3:27–32

Crawford D, Amstad P, Yin Foo D, Cerutti P (1989) Constitutive and phorbolmyristate acetate regulated antioxidant defense of mouse epidermal JB6 cells. Mol Carcinog 2:136–143

Creasia D, Nettesheim P (1974) Particulates and chronic inflammation increase rates of carcinogenesis. In: Karbe E, Park J (eds) Experimental lung cancer. Springer, Berlin Heidelberg New York

De Fromentel C, Sossi T (1992) The p53 tumor suppressor gene: a model for investigating human mutagenesis. Genes Chrom Cancer 4:1-4

Devary Y, Gottlieb R, Lau L, Karin M (1991) Rapid preferential activation of the c-jun gene during the mammalian UV response. Mol Cell Biol 11:2804-2811

Devary Y, Gottlieb R, Smeal T, Karin M (1992) The mammalian ultraviolet response is triggered by activation of src tyrosine kinases. Cell 71:1081–1091

Edginton T, Ross R, Silverstein S (eds) (1987) Perspectives in inflammation, neoplasia and vascular cell biology. Wiley, New York

Elroy-Stein O, Bernstein Y, Groner Y (1986) Overproduction of human Cu-Zn-superoxide dismutase in transfected cells: extenuation of paraquat-mediated cytotoxicity and enhancement of lipid peroxidation. EMBO J 5:615–622

Epstein J (1978) Photocarcinogenesis: a review. Natl Cancer Inst Monogr 50:13–25

Feng J, Villeponteau B (1990) Serum-stimulation of the c-fos enhancer induces reversible changes in c-fos chromatin structure. Mol Cell Biol 10:1126–1133

Feng J, Villeponteau B (1992) High resolution analysis of c-fos chromatin accessibility using a novel DNaseI-PCR assay. Biochim Biophys Acta 1130:253–258

Forbes P, Davies R, Urbach F (1978) Experimental ultraviolet carcinogenesis: wavelength interactions and time-dose relationships. Natl Cancer Inst Monogr 50:31-38

Gey F (1987) Plasma levels of antioxidant vitamins in relation to ischemic heart disease and cancer. Am J Nutr 45:1368–1377

Ghosh R, Amstad P, Cerutti P (1993) UVB-induced DNA breaks interfere with transcriptional inductin of c-fos. Mol Cell Biol 13:6992–6999

Gindhart T, Nakamura Y, Stevens L, Hegameyer G, West M, Smith B, Colburn N (1985) Genes and signal transduction in tumor promotion. Conclusions from studies with promoter resistant variants of JB6 mouse epidermal cells. In: Mass M, Kaufman D, Siegfried J, Steele V, Nesnow S (eds) Cancer of the respiratory tract. Raven, New York

Hariharan P, Cerutti P (1977) Formation of products of the 5,6-dihydroxy-dihydrothymine-type by ultraviolet light in HeLa cells. Biochemistry 16:2791–2795

Hirschi M, Netrawali M, Remsen J, Cerutti P (1981) Formation of DNA single strand breaks by near-ultraviolet and gamma-rays in normal and Bloom's syndrome skin fibroblasts. Cancer Res 41:2003–2007

Hodgson E, Fridovich I (1975) The interaction of bovine erythrocyte superoxide dismutase with hydrogen peroxide: chemiluminescence and peroxidation. Biochemistry 14:5299–5303

Hollstein M, Sidransky D, Vogelstein B, Harris C (1991) p53 Mutations in human cancers. Science 253:49-53

Kaina B, Stein B, Schönthal A, Rahmsdorf H, Ponta H, Herrlich P (1990) An update of the mammalian UV response: gene regulation and induction of a protective function. Life Sci 182:652–654

Kensler T, Bush D, Kozumbo W (1983) Inhibition of tumor promotion by a biomimetic superoxide dismutase. Science 221:75–77

Kozumbo W, Trush M, Kensler T (1985) Are free radicals involved in tumor promotion. Chem Biol Interact 54:199-207

Lewis J, Hamilton T, Adams D (1986) The effect of macrophage development on the release of reactive oxygen intermediates and lipid oxidation products, and their ability to induce oxidative DNA damage in mammalian cells. Carcinogenesis 7:813–818

Mao G, Thomas P, Lopaschuk G, Poznansky M (1983) Superoxide dismutase (SOD)-catalase conjugates. J Biol Chem 268:416–420

Mass M, Kaufman D, Siegfried J, Steele V, Nesnow S (eds) (1985) Cancer of the respiratory tract: predisposing factors. Raven, New York

Menkes M, Comstock G, Vuilleumier J, Helsing K, Rider A, Brookmeyer R (1986) Serum beta-carotene, vitamins A and E, selenium and the risk of lung cancer. N Engl J Med 315:1250–1254

Muehlematter D, Larsson R, Cerutti P (1988) Active oxygen induced DNA strand breakage and polyADP-ribosylationin promotable and non-promotable JB6 mouse epidermal cells. Carcinogenesis 9:239–245

Niggli H, Cerutti P (1983) Cyclobutane-type pyrimidine photodimer formation and excision in human skin fibroblasts after irradiation with 313 nm ultraviolet light. Biochemistry 22:1390–1395

Ochi T, Cerutti P (1987) Clastogenic action of hydroperoxy-5,8,11,13-icosatetraenoic acids on the mouse embryo fibroblasts C3H/10T1/2. Proc Natl Acad Sci 84:990–994

O'Connell J, Klein-Szanto A, DiGiovanni D, Fries J, Slaga T (1986) Enhanced malignant progression of mouse skin tumors by the free-radical generator benzoyl peroxide. Cancer Res 46:2863–2865

Omar B, McCord J (1990) The cardioprotective effect of Mn-superoxide dismutase is lost at high doses in the postischemic isolated rabbit heart. Free Radic Biol Med 9:473–478

Pourzand C, Cerutti P (1993) Genotypic mutation analysis by RFLP/PCR. Mutat Res 288:113–121

Radler-Pohl A, Sachsenmaier C, Gebel S, Auer HP, Bruder J, Rapp U, Angel P, Rahmsdorf H, Herrlich P (1993) UV-induced activation of AP-1 involves obligatory extranuclear steps including Raf-1 kinase. EMBO J 12:1005–1012

Ronai Z, Okin E, Weinstein IB (1988) Oncogene 2:204–210

Scott M, Meshnik S, Eaton J (1987) Ultraviolet light induces the expression of oncogenes in rat fibroblasts and human keratinocytes. Biol Chem 262:3640–3645

Shah G, Ghosh R, Amstad P, Cerutti P (1993) Mechanism of induction of c-fos by ultraviolet B (290320 nm) in mouse JB6 epidermal cells. Cancer Res 53:38–45

Shibanuma M, Kuroki T, Nose K (1988) Induction of DNA replication and expression of protooncogenes c-myc and c-fos in quiescent Balb/3T3 cells by xanthine/xanthine oxidase. Oncogene 3:17–21

Slaga T, Solanki V, Logani M (1983) Studies on the mechanisms of action of antitumor promoting agents: suggestive evidence for the involvement of free radicals in promotion. In: Nygaard O, Simic M (eds) Radioprotectors and anticarcinogens. Academic, New York

Stalder J, Groudine M, Dodgson J, Engel J, Weintraub H (1980) Hemoglobin switching in chickens. Cell 19:973:980

Viaje A, Slaga T, Wigler M, Weinstein I (1977) Effects of antiinflammatory agents on mouse skin tumor promotion, epidermal DNA synthesis, phorbol ester-induced cellular proliferation, and production of plasminogen activator. Cancer Res 37:1530-1536

Weintraub H (1987) Assembly and propagation of repressed and depressed chromosomal states. Cell 42:705–711

Zimmerman R, Cerutti P (1984) Active oxygen acts as a promoter of transformation in mouse embryo fibroblasts. Proc Natl Acad Sci USA 81:2085–2087

3 DNA Damage by Free Radicals. Mechanism, Meaning and Measurement

B. Halliwell

3.1 Introduction

It is well established that aerobes constantly produce small amounts of oxygen-derived species, such as superoxide radical ($O_2^{\bullet-}$), hydrogen peroxide (H_2O_2), and hypochlorous acid (HOCl), the latter being generated by the enzyme myeloperoxidase in neutrophils (for reviews, see

Table 1. Methods used to subject cells to oxidative stress that has produced increased intracellular DNA damage

---

Elevated O_2 concentrations
Exposure to activated phagocytic cells
Exposure to "redox cycling" drugs (e.g. alloxan, paraquat, menadione)
Exposure to cigarette smoke
Exposure to ozone
Exposure to ionizing radiation
Direct addition of hydrogen peroxide or organic peroxides
Exposure to "autoxidizing" agents (e.g. dihydroxyfumarate, pyrogallol, adrenalin)
Exposure to xanthine oxidase[a] plus its substrates (xanthine, hypoxanthine)
Addition of tumor necrosis factor

See Halliwell and Aruoma (1991).
[a] Care must be taken in use of commercial xanthine oxidase, which is often heavily contaminated with proteases and other material directly injurious to cells.

di Guiseppi and Fridovich 1984; Halliwell and Gutteridge 1989; Weiss 1989). Exposure of living organism to background levels of ionizing radiation leads to homolytic fission of oxygen–hydrogen bonds in water to produce highly reactive hydroxyl radicals, $OH^\bullet$ (for review, see von Sonntag 1987). Hydroxyl radicals can also be generated when H_2O_2 comes into contact with certain transition metal ion chelates, especially those of iron and copper. In general, the reduced forms of these metal ions (Fe^{2+}, Cu^+) produce $OH^\bullet$ at a faster rate upon reaction with H_2O_2 than the oxidized forms (Fe^{3+}, Cu^{2+}), and so reducing agents such as $O_2^{\bullet-}$ and ascorbic acid can often accelerate $OH^\bullet$ generation by metal ion/H_2O_2 mixtures (Halliwell and Gutteridge 1990a). Another potentially physiologically important source of $OH^\bullet$ is the interaction of nitric oxide radical ($NO^\bullet$) with $O_2^{\bullet-}$ (Beckman et al. 1990).

Aerobes have evolved antioxidant defenses to protect themselves against the oxygen-derived species generated in vivo. These defenses include enzymes (such as superoxide dismutases, catalase, and glutathione peroxidases), low-molecular-mass agents (such as α-tocopherol and ascorbic acid), and proteins that bind metal ions in forms unable to accelerate free radical reactions (Sies 1991; di Guiseppi and Fridovich 1984; Halliwell and Gutteridge 1990a,b). Oxidative stress results when oxygen-derived species are not adequately removed. This can happen if antioxidants are depleted and/or if the formation of oxygen-derived species is increased beyond the ability of the defenses to cope with them (Sies 1991).

Subjecting cells to oxidative stress can result in severe metabolic dysfunctions, including peroxidation of membrane lipids, depletion of nicotinamide nucleotides, rises in intracellular free Ca^{2+} ions, cytoskeletal disruption, and DNA damage. The latter is often measured as formation of single-strand breaks, double-strand breaks, or chromosomal aberrations. Indeed, DNA damage has been almost invariably observed in a wide range of mammalian cell types exposed to oxidative stress in a number of different ways (Table 1). Oxidative stress (Larrick and Wright 1990) and DNA damage (Zimmerman et al. 1989) also occur when some mammalian cells are exposed to tumor necrosis factors. Oxidative stress may additionally play some role in the carcinogenicity of several compounds (Floyd 1990; Frenkel 1992), including asbestos cigarette smoke (Leanderson and Tagesson 1989, 1992; Loft et al. 1992; Kiyosawa et al. 1990), nitroso compounds (Kelly et al. 1992; Chung and Xu 1992) and certain metals, such as nickel (Kasprzak 1991; Kasprzak et al. 1992). There is also considerable interest in the relationship between oxidative stress and nongenotoxic carcinogenesis.

3.2 Possible Mechanisms of DNA Damage Induced by Oxidative Stress

Why does oxidative stress cause DNA damage? In the case of externally generated oxygen-derived species (e.g., when cells are incubated with H_2O_2, activated phagocytes, or xanthine oxidase plus its substrates), damage is usually inhibited by adding catalase, showing that H_2O_2 is needed. Superoxide dismutase (SOD) does not usually inhibit much, which could mean either that $O_2^{\bullet-}$ is not involved in the DNA damage or that SOD does not enter cells easily. That the latter interpretation is correct in at least some cell systems is shown by several observations that SOD can protect cells from the toxicity of H_2O_2 or organic hydroperoxidases under conditions where SOD can enter the cells (e.g., Kyle et al. 1988; Nakae et al. 1990).

It has been known for many years that neither $O_2^{\bullet-}$ nor H_2O_2 causes any strand breakage in DNA, if the reaction mixture is carefully freed of transition metal ions (see, for example, Rowley and Halliwell 1983).

Our later work (Aruoma et al. 1989a,b, 1991) confirmed this inability of $O2^{\bullet-}$ or H_2O_2 at physiologically relevant concentrations to damage DNA, by looking for chemical changes in the purine and pyrimidine bases (see below). Hence, DNA damage by oxidative stress in vivo is unlikely to involve direct attack of $O2^{\bullet-}$ or H_2O_2 upon the DNA.

Two explanations of the DNA damage have been advanced (Fig. 1). First, it is possible that the damage is due to $OH^{\bullet}$ radical formation, a proposal first clearly stated by Mello-Filho et al. (1984). Thus, it is envisaged that H_2O_2, which crosses biological membranes easily, can penetrate to the nucleus and react with iron and/or copper ions to form $OH^{\bullet}$. Because of the high reactivity of $OH^{\bullet}$ and its resultant inability to diffuse significant distances within the cell (Halliwell and Gutteridge 1990a), this mechanism is only feasible if the $OH^{\bullet}$ is generated from H_2O_2 by reaction with metal ions bound upon or very close to the DNA. One possibility is that these metal ions are always present bound to the DNA in vivo. For example, copper ions are thought to be present in chromosomes (Lewis and Laemmli 1982; Dijkwel and Wenink 1986; Prutz et al. 1990). Copper ions bind to guanine residues in DNA (Geier-stanger et al. 1991) and are very effective in promoting H_2O_2-dependent damage to isolated DNA and to DNA within chromatin in vitro (see below). A second possibility (suggested by Halliwell 1987) is that the metal ions are released within the cell as a result of oxidative stress and then bind to the DNA. Thus, just as oxidative stress causes rises in intracellular free Ca^{2+} (Orrenius et al. 1989), it may cause rises in intracellular free iron and/or copper ions by interfering with normal intracellular sequestration mechanisms. Some of these released ions may then bind to DNA and make it a target for oxidative damage.

A second explanation of the ability of oxidative stress to cause DNA damage is that the stress triggers a series of metabolic events within the cell that lead to activation of nuclease enzymes, which cleave the DNA backbone. Oxidative stress causes rises in intracellular free Ca^{2+}, which can fragment DNA by activating Ca^{2+}-dependent endonucleases (Orre-nius et al. 1989; Farber 1990; Ueda and Shah 1992) in a mechanism with some of the features of apoptosis (see Wyllie 1980). An example of apoptosis is the killing of immature thymocytes by glucocorticoid hor-mones, which activate a cell-destructive process that apparently in-volves DNA fragmentation by a Ca^{2+}-dependent nuclease.

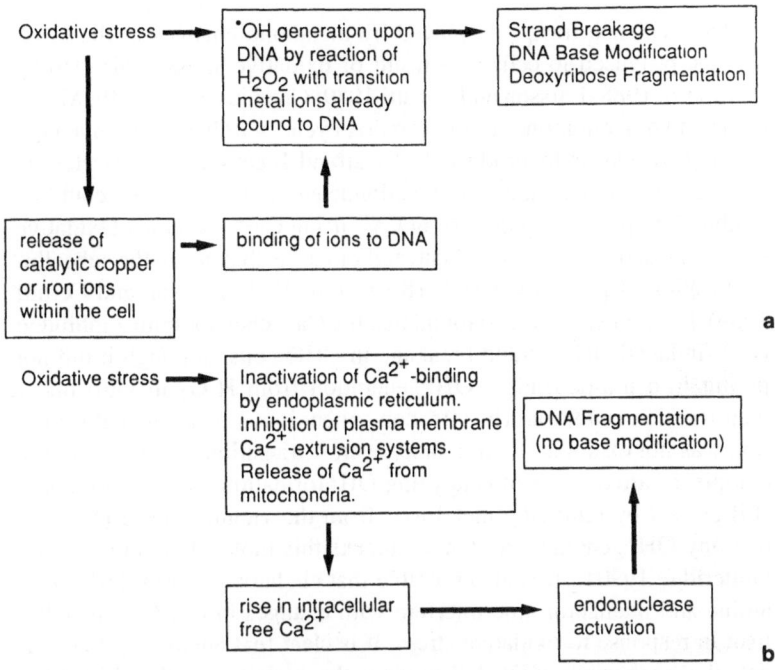

Fig. 1a,b. Hypotheses to explain DNA damage resulting from exposing cells to oxidative stress. **a** Fenton chemistry; **b** nuclease activation

These two mechanisms (DNA damage by OH• or by activation of nucleases) are not mutually exclusive, i.e., they could both take place (Fig. 1). Indeed, there is evidence consistent with the existence of both mechanisms. Their relative importance may depend on the cell type used and on how the oxidative stress is imposed (Halliwell and Aruoma 1991). For example, chelating agents that bind ions into chelates unable to generate OH• (such as desferrioxamine, desferrithiocin, and phenanthroline) can often protect cells against DNA damage and other toxic effects of oxidative stress (Mello-Filho et al. 1984; Imlay and Linn 1988; Halliwell and Aruoma 1991). The effects of desferrioxamine are variable, since in general it does not cross cell membranes readily, although it appears to enter some cell types (such as hepatocytes) more readily than it enters others.

The evidence for a role played by metabolic changes in the DNA damage produced in cells as a result of oxidative stress is also strong (Birnboim 1988; Larsson and Cerutti 1989; Orrenius et al. 1989). Menadione and other quinones (which "redox cycle" within cells to give $O_2^{\bullet -}$ and H_2O_2) appear to produce DNA strand breaks in hepatocytes by Ca^{2+}-dependent activation of an endonuclease. DNA damage could be inhibited by preventing the rise in Ca^{2+} using Ca^{2+} chelators. Oxidative stress can also sometimes activate and/or cause changes in the subcellular location of protein kinase C (Kass et al. 1989; Larsson and Cerutti 1989). Cantoni et al. (1989) found that the Ca^{2+} chelator quin 2 inhibited H_2O_2-induced DNA strand breakage in CHO cells, although it did not inhibit iron ion-dependent $OH^{\bullet}$-generation from H_2O_2 in vitro under their reaction conditions (its effect on copper ion-dependent $OH^{\bullet}$-formation was not examined). Of course, even if transition metal ion–quin 2 complexes are capable of catalyzing $OH^{\bullet}$-formation, the chelator could still protect by removing metal ions from the vicinity of the DNA, so that any $OH^{\bullet}$-generated no longer attacks this molecule (Halliwell and Gutteridge 1990a). It is also possible that chelators such as desferrioxamine and phenanthroline interfere with changes in cell Ca^{2+} metabolism in response to oxidative stress. It is clear that attempting to elucidate the mechanism of DNA damage in the nucleus of cells subjected to oxidative stress by adding free radical scavengers or metal ion chelators to the surrounding media is unlikely to give unambiguous answers.

3.3 The Physiological Importance of DNA Damage Induced by Oxidative Stress

Why is it important to understand the mechanism of DNA damage by oxidative stress? Oxidative stress, imposed by a variety of mechanisms (including increased O_2 concentrations), has been convincingly shown to be mutagenic to bacterial and mammalian cells (reviewed by Halliwell and Aruoma 1991; also see Essigmann and Wood 1993; Feig and Loeb 1993). For example, *Escherichia coli* mutants lacking SOD activity show greatly enhanced rates of spontaneous mutation (Touati 1989; Prieto-Alamo et al. 1993). Moraes et al. (1990) studied the pattern of mutations obtained in a gene of a shuttle plasmid when simian cells transfected with this plasmid were exposed to H_2O_2. Both single base changes and deletions were observed. The majority of base changes were at GC base pairs, the GC → AT base transition being predominant. Treatment of the plasmid with H_2O_2 in vitro before transfection did not produce an increased number of mutations (unless iron ions were added), consistent with the inability of H_2O_2 to react directly with DNA. McBride et al. (1991) found similar results when single-stranded M13mp2 DNA was incubated with Fe^{2+} ions under aerobic conditions and then transfected into *E. coli*. Cheng et al. (1992) showed that 8-hydroxydeoxyguanosine residues in DNA can lead to G → T and A → C substitutions.

Mutations induced by oxidative stress may lead to cancer. Ionizing radiation is well known to be both mutagenic and carcinogenic. Since much of the cell damage caused by such radiation involves OH• production by homolytic fission of the oxygen–hydrogen bonds in water, OH• can probably be classified as a complete carcinogen. Base pair changes and some frameshifts are the most common mutations observed in cells exposed to ionizing radiation (for review, see Breimer 1988, 1990). Chemical changes in the DNA bases, single- and double-strand breaks, and enhanced expression of certain proto-oncogenes have also been described (von Sonntag 1987). However, the precise relationship between these different events and the development of cancer is uncertain. Thus, the chemical changes in DNA may themselves somehow lead to cancer (for discussion, see Floyd 1990). An unrepaired lesion in DNA may be bypassed in an error-prone fashion. Resynthesis of DNA after excision repair may conceivably introduce errors.

There are many steps between a healthy cell and a malignant tumor (Trush and Kensler 1991; Frebourg and Friend 1992). Cancer biologists have often referred to at least three stages: initiation (an irreversible change in DNA), promotion (probably involving changes in gene expression), and progression (further changes in DNA leading to the eventual production of a malignant tumor), although these distinctions have become less clear-cut in recent years. Both Zimmerman and Cerutti (1984) and Weitzman et al. (1985) showed that a clone of C3H mouse fibroblasts exposed to activated human neutrophils or to hypoxanthine plus xanthine oxidase underwent malignant transformation. Nassi-Calo et al. (1989) showed that H_2O_2 also transformed these cells, an action prevented by the chelating agent *o*-phenanthroline. The ability of oxidative stress to induce transformation has also been shown in human lung fibroblasts (Weitberg and Corvese 1990). Weitzman et al. (1988) reported that DNA isolated from C3H mouse fibroblasts that had been transformed by exposure to activated neutrophils could sometimes transform another cell line, NIH-3T3, when the DNA was transfected into the latter cells.

Although most attention has been paid in the literature to the action of oxygen-derived species as promoters of carcinogenesis, their ability to damage DNA and produce alterations in gene expression implies that they could be involved in all stages of carcinogenesis (Trush and Kensler 1991). Indeed, it has been argued (see, for example, Totter 1980; Ames 1989) that continuous oxidative damage to DNA by free radical mechanisms is a significant cause of cancer in humans and explains why cancer incidence increases sharply with age. Of course, DNA damage resulting from oxidative stress (or from any other mechanism; Oller and Thilly 1992) need not necessarily lead to cancer. Low levels of damage may be efficiently repaired with a minimal risk of error (Breimer 1991; Lindahl 1993). High levels of oxidative stress may lead to cell death, so that initiated cells do not remain in the organism. Thus, an intermediate level of oxidative stress is most likely to predispose to malignancy. Another relevant observation may be the ability of low level oxidative stress to stimulate cell proliferation (Burdon and Rice-Evans 1989; Meier et al. 1990; Goligorsky et al. 1992; Murrell et al. 1990).

It is interesting to note the association of chronic inflammation (involving phagocytic production of $O_2^{\bullet-}$ and H_2O_2) with malignancy in such human diseases as ulcerative colitis, Crohn's disease, and reflux esophagitis (Weitzman and Gordon 1990). Cerutti et al. (1989) showed that one difference between a clone of mouse epidermal cells that was promotable by xanthine/xanthine oxidase and a nonpromotable clone was that the latter had lower levels of SOD and catalase and was more sensitive to killing by oxygen-derived species. Thus, increased antioxidant defenses, by protecting against cell death resulting from oxidative stress, may sometimes conceivably, and ironically, lead to increased cancer.

Another interesting observation is that exposure of DNA to OH$^{\bullet}$ can render the DNA antigenic, an observation perhaps relevant to the formation of anti-DNA antibodies in some human chronic inflammatory diseases (Blount et al. 1992; Alam et al. 1993).

3.4 Probing the Mechanism of DNA Damage
in Cells Exposed to Oxidative Stress: The Principles

We have already commented that it is difficult to gain information about
the mechanism of oxidative stress-induced DNA damage by using antioxidants and scavengers. Another means of implicating free radicals as
damaging agents is to use what I have called a "finger-print" approach:
if a free radical produces a unique pattern of chemical change in a
biological molecule, then observation of the same pattern in vivo is
evidence that the radical attacked that molecule. The damage pattern
must be unique to that radical. Therefore, we and others (e.g., Carmichael et al. 1992; Epe et al. 1988, 1993) set out to characterize the
chemical changes produced in DNA by different oxygen-derived
species. We used the technique of gas chromatography/mass spectrometry with selected ion monitoring, largely developed for work with
DNA by Dizdaroglu (reviewed in 1991). We concentrated on damage to
the purine (adenine, guanine) and pyrimidine (cytosine, thymine) bases,
because sugar-damage products are much less chemically distinctive
(von Sonntag 1987). We found, as expected, that $O_2^{\bullet-}$ and H_2O_2 at
physiologically relevant concentrations do not themselves cause any
base damage in DNA (Aruoma et al. 1989a,b, 1991).   ·

Several studies (for reviews see Steenken 1989; Dizdaroglu 1991)
had already shown OH• reacts in a multiplicity of ways with all four
DNA bases. Thus, OH• can add on to guanine residues at C4, C5, and
C8 positions to give hydroxyguanine radicals that can have various
fates. For example, addition of OH• to C8 of guanine produces a radical
that can be reduced to 8-hydroxy-7,8-dihydroguanine, oxidized to 8-hy-
droxyguanine (8-OH-Gua), or can undergo ring opening followed by
one electron reduction and protonation to give 2,6-diamino-4-hydroxy-
5-formamidopyrimidine, usually abbreviated as FapyGua. Figure 2
shows the structures of some of these products. Similarly, OH• can add
on to C4, C5, or C8 of adenine. Among other fates, the C8 OH• adenine
radical can be converted into 8-hydroxyadenine (8-OH-Ade) by oxidation or can undergo ring opening followed by one-electron reduction to
give 5-formamido-4,6-diaminopyrimidine (FapyAde). Pyrimidines are
also attacked by OH• to give multiple products. Thus, thymine can form
cis and trans thymine glycols (5,6-dihydroxy-6-hydrothymines), 5-hy-
droxy-5-methylhydantoin, 5,6-dihydrothymine, and 5-hydroxymethy-

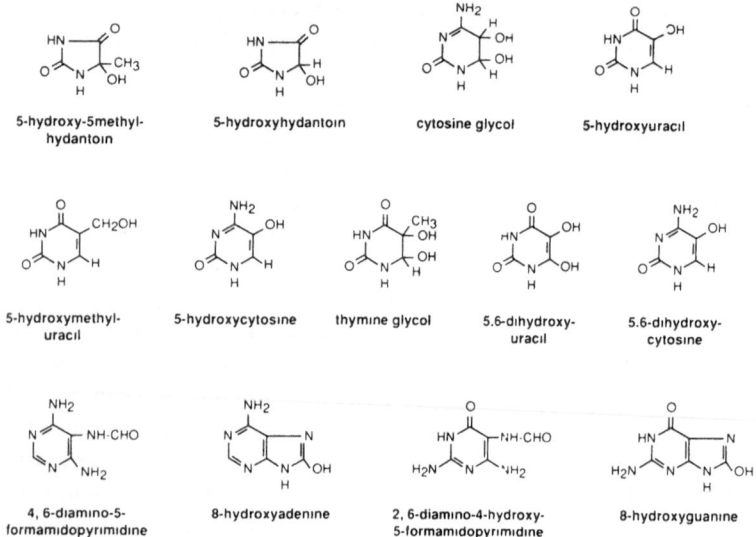

Fig. 2. Some of the end products that result from attack of hydroxyl radicals upon the bases of DNA

luracil. Cytosine can form several products, including cytosine glycol and 5,6-dihydroxycytosine (Fig. 2; for reviews, see Dizdaroglu 1991; Steenken 1989). In addition, OH• generation within whole cells or isolated chromatin can result in formation of cross-links between DNA bases and amino acid residues in nuclear proteins. Thus, thymine–tyrosine, thymine–aliphatic amino acid, and cytosine–tyrosine links have been identified in isolated calf thymus chromatin subjected to γ-irradiation or treated with metal ions and H_2O_2 (Dizdaroglu 1991; Nackerdien et al. 1991).

Molecular biologists have examined the likely physiological effects of these various lesions in DNA. 8-OH-Gua (and, perhaps 8-OH-Ade) may lead to mutations by inducing misreading of the base itself and possibly of the adjacent bases (Breimer 1991; Shibutani et al. 1991; Tchou and Grollman 1993; Retel et al. 1993). Thymine glycol may have some mutagenic action, and it can be lethal if not removed from the DNA before replication (Evans et al. 1993). Ring-fragmented bases, as well as thymine glycol, can block DNA replication (Evans et al. 1993).

Abasic sites, which can also result from attack of OH•, can be mutagenic as well (Breimer 1991).

It is clear that OH• produces multiple changes in DNA, whereas O_2•⁻ and H_2O_2 have no effect. What about other oxygen-derived species? Single oxygen is able to produce limited strand breakage in isolated DNA, and its ability to modify the DNA bases is also limited. Thus, Dizdaroglu (personal communication) and Sies (1993) found small amounts of 8-OH-Gua and FapyGua but no other significant base changes in DNA exposed to singlet O_2 generated by the thermal decomposition of an endoperoxide. Thus, singlet O_2 seems specific for guanine and does not induce the extensive pattern of DNA base modification produced by OH•. Peroxyl and alkoxyl radicals might preferentially react with guanine, whereas HOCl would be expected to chlorinate ⁻NH_2 groups. Oxides of nitrogen may deaminate DNA bases, whereas NO• might react with O_2•⁻ to give OH•·(Wink et al. 1991; Beckman et al. 1990). More work is required to characterize these reactions in detail, but I doubt that any species other than OH• produces the extensive pattern of base modification shown in Fig. 2. Several authors have reported that peroxidizing lipids damage DNA but, in interpreting the data, one must bear in mind that peroxidizing lipids produce a range of reactive oxygen species including OH•, H_2O_2, singlet oxygen, peroxyl radicals, and alkoxyl radicals (Gutteridge and Halliwell 1990), and the exact contributions of these species to the DNA damage observed need to be determined. Lipid peroxides also decompose at body temperature in the presence of transition metal ions to give a huge range of products, including carbonyl compounds, such as malondialdehyde (MDA) and the unsaturated aldehyde 4-hydroxy-2-*trans*-nonenal, which has been shown to be mutagenic to mammalian cells (Canonero et al. 1990). If these aldehydes are generated in the vicinity of DNA, they may be able to combine with it to form distinctive products. Thus, MDA reacts with adenine, cytosine, and guanine (Stone et al. 1990a,b), and a guanine–MDA adduct has been identified in human urine (Hadley and Draper 1990). The product of reaction of hydroxynonenal with deoxyguanosine has also been characterized (Sodum and Chung 1988).

Humans are constantly exposed to background levels of ionizing radiation, which will generate some OH• in vivo. This radical may also arise by reaction of metal ions with H_2O_2 in vivo. Thus, it is not surprising to find that repair systems have evolved to remove at least

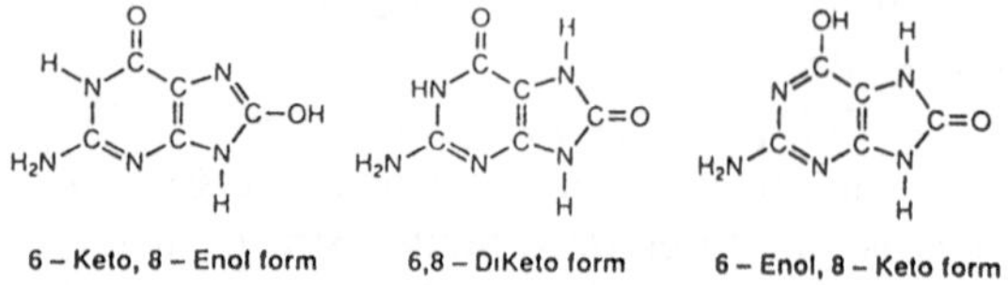

6 – Keto, 8 – Enol form   6,8 – DiKeto form   6 – Enol, 8 – Keto form

Fig. 3. Keto-enol tautomerism of 8-hydroxyguanine

some of the lesions in DNA that can result from attack of OH• (for review, see Breimer 1991).

Some of the modified DNA bases shown in Fig. 2, and their nucleosides (base-deoxyribose), have been detected in the urine of humans and other mammals. Thus, 8-OH-Ade, 7-methyl-8-hydroxyguanine, thymine glycol, thymidine glycol, hydroxymethyluracil, 8-OH-Gua, and 8-hydroxydeoxyguanosine have been detected in mammalian urine (Ames 1989; Stillwell et al. 1989; Fraga et al. 1990). The presence of these products in urine suggests that oxidative damage to the DNA bases does occur in vivo and that repair systems are active to cleave modified bases from DNA. However, it is possible that some of the modified bases excreted originate from the diet or from the metabolism of gut flora. In addition, it is possible that DNA released from dead and dying cells within an organism undergoes rapid oxidative damage (since cell disruption can increase free radical reactions; see Halliwell and Gutteridge 1984). 8-hydroxydeoxyguanosine may arise from dGTP that has become oxidized in vivo: the damaged dGTP may be removed before incorporation into DNA, in a process that has been called "sanitization of the nucleotide pool" (Mo et al. 1992).

Hence, one must be cautious in using the amounts of modified DNA bases excreted from the body as an index of the extent of repair of oxidative DNA damage in healthy cells.

3.5 Methods of Measuring 8-Hydroxyguanine and 8-Hydroxydeoxyguanosine

I should first clarify a point of nomenclature. 8-hydroxyguanine (8-OHG) describes the purine base guanine in which the hydrogen atom at position 8 is replaced by an ⁻OH group (Fig. 2). It can be released from

DNA by acidic hydrolysis. If enzymic hydrolysis is used instead, 8-OHG may be released still attached to the 2-deoxyribose sugar. This product is called 8-hydroxy-2'-deoxyguanosine (8-OHdG). Guanine is the base, guanosine the nucleoside (basedeoxyribose). In fact 8-OHG should more properly be called 8-oxo-7-hydroguanin, since the keto-enol tautomerism favors the 6,8-diketo form (Fig. 3; see Aida and Nishimura 1987; Oda et al. 1991). However, the "standard" nomenclature will be used in this article.

3.5.1 High Performance Liquid Chromatography

The development of a high performance liquid chromatography (HPLC)-based technique, coupled with highly sensitive electrochemical detection, for the measurement of 8-OHdG released from DNA by enzymic digestion was the main impetus that has led to the popularity of this product for measurement as a putative index of oxidative DNA damage (Kasai and Nishimura 1984; Kasai et al. 1984, 1986; Floyd et al. 1986), although HPLC techniques for other bases have been described (e.g., Kojima et al. 1989; Berger et al. 1990). The studies of Aruoma et al. (1989a,b, 1991) and others helped to validate the choice of 8-OHdG by showing that 8-OHG is a major product of damage to DNA by biologically relevant oxygen-derived species such as OH$^\bullet$ and $^1O_2$, i.e., if one had to pick a single product to measure, 8-OHG would be the one to choose.

HPLC-based measurement of 8-OHdG (Floyd et al. 1986) has rapidly become popular as a means of gaining some information about damage to DNA in intact cells and whole organisms. For example, the amount of 8-OHdG in the DNA of certain subpopulations of rat liver mitochondria was found to be considerably higher than in rat liver nuclear DNA, leading to proposals about the role of mitochondrial DNA damage in aging and carcinogenesis (Richter 1988; Richter et al. 1988). The steady state level of mitochondrial 8-OHdG in human heart has been reported to rise with age (Hayakawa et al. 1992). Exposure of numerous cell types to oxidative stress has been reported to increase the 8-OHdG content of their DNA (reviewed by Ames 1989; Floyd 1990). Oxidative DNA damage (as 8-OHdG) has been measured in human sperm and the amount detected shown to increase in subjects with low intakes of ascorbic acid (Fraga et al. 1991).

These pioneering measurements of 8-OHdG have produced evidence that oxidative damage to DNA does occur in vivo, although care must be used in interpreting the data. One must be especially cautious in attempting to use levels of 8-OHdG (or of any other single product) as a quantitative measure of DNA base damage by oxygen-derived species such as OH• (Halliwell and Aruoma 1991). When OH• attacks DNA bases, radicals are formed that have different fates, depending on experimental conditions. For example, attack of OH• on guanine can lead to formation of 8-OHG by oxidation of the C8 OH adduct radical, but this radical can lead to other products as well (such as FapyGua). Hence, different amounts of 8-OHG can result from attack of the same amount of OH• on guanine in DNA. This is one reason why changes in 8-OHdG levels do not necessarily reflect changes in the amount of free radical attack on DNA (Halliwell and Aruoma 1991). To take some examples, iron ion-dependent systems generating OH• cause substantial formation of both FapyGua and 8-OHG in DNA (Aruoma et al. 1989a,b), whereas systems containing copper ions and H_2O_2 greatly favor 8-OHG over FapyGua (Aruoma et al. 1991; Dizdaroglu et al. 1991a,b). When mammalian chromatin is isolated, suspended in aqueous buffer solution and exposed to radiation-generated OH·, the relative amounts of 8-hydroxy-purines and formamidopyrimidines generated depend on the environment provided by gases used to saturate the aqueous solution. For example, anoxic conditions favor formamidopyrimidines over 8-hydroxypurines (Gajewski et al. 1990).

Thus, HPLC-based analysis of 8-OHdG as a method of measuring oxidative DNA damage, despite its undoubted value, has intrinsic limitations. Another problem is the frequency with which coeluting peaks occur when HPLC is applied to complex biological fluids: a peak must never be assumed to represent 8-OHdG merely on the basis of its retention time. Electrochemical behavior and (if possible) absorbance of fluorescence spectra should be checked for identity with those of authentic 8-OHdG. Mass spectrometry is the technique of choice for unequivocal identification, if sufficient material is available.

One must also be careful in interpreting the results of such measurements. An increase in the 8-OHdG content of DNA in a cell might mean increased oxidative damage, but decreased repair could also explain it.

Thus, do mitochondria repair oxidized DNA more slowly than the nucleus does and so show higher levels of 8-OHG (Richter et al. 1988)? Another problem to be considered is that of dead cells. Cells die constantly in the human body. Dead, disrupted cells are known to undergo lipid peroxidation faster than healthy cells (Halliwell and Gutteridge 1984): perhaps they also undergo oxidative DNA damage faster as well, so that excretion of DNA base damage products is not necessarily a reflection of the extent of oxidative DNA damage in healthy cells. Food often contains oxidized lipids generated during storage and cooking – perhaps DNA is also damaged. The damaged DNA could be hydrolyzed in the intestine, and the damaged base products absorbed and eventually excreted. An alternative explanation of the results of Fraga et al. (1991) would be that ascorbate-depleted sperm die faster and then their DNA oxidizes.

In addition, interpretation of measurements of base damage products excreted in urine as an index of "whole body oxidative DNA damage" presupposes that these products are not significantly metabolized after cleavage from the DNA. Knowing the tremendous capacity of mammals to metabolize "foreign compounds," this seems unlikely. Thus enzymes exist that can deaminate DNA bases: perhaps they can also metabolize modified bases (Agarwell and Parks 1978; Glantz and Lewis 1978; Geborek et al. 1992). These areas remain to be explored.

3.5.2 Gas Chromatography–Mass Spectrometry

Characterization of various types of damage to DNA by oxygen-derived species can be achieved by the technique of gas chromatography–mass spectrometry (GC–MS), which may be applied to DNA itself or to DNA–protein complexes such as chromatin (Dizdaroglu 1991). For GC–MS, the DNA or chromatin is hydrolyzed (usually by heating with formic acid) and the products are converted to volatile derivatives, which are separated by GC and conclusively identified by the structural evidence provided by a mass spectrometer. Stable isotope-labeled bases may be used as internal standards (Dizdaroglu 1993). High sensitivity and selectivity of detection can be achieved by operating the mass spectrometer in the selected ion monitoring (SIM) mode. In SIM, the mass spectrometer is set to monitor several ions derived by fragmentation of a particular product during the time at which this product is expected to emerge from the GC column. The GC–MS–SIM technique can be used to study the precise mechanism by which DNA is damaged in cells subjected to oxidative stress. Thus, if damage is due to OH• generation, then the whole spectrum of base products characteristic of OH• attack should be detected, as has been observed in chromatin isolated from γ-irradiated human cells in culture (Nackerdien et al. 1992), from murine hybridoma cells after treatment with H_2O_2 (Dizdaroglu et al. 1991b), and in DNA isolated from primate tracheal epithelial cells exposed to ozone or to cigarette smoke (Halliwell and Aruoma 1991). If, for example, singlet O_2 were responsible for the DNA damage, then a much more limited range of products should be measurable, such as 8-OHG and FapyGua (Floyd et al. 1989; Devasagayam et al. 1991). The results of GC–MS analysis of modified bases in DNA have usually been expressed as nonomoles of modified bases per milligram of DNA (equivalent to pmol/μg DNA). However, it is easy to convert these data into the actual number of bases modified. Dividing the figure of nmol bases/mg DNA by 3.14 (or multiplying by 0.318) gives the number of modified bases per 10^3 bases in DNA, i.e., 1 nmol/mg DNA corresponds to about 318 modified bases per 10^6 DNA bases.

3.5.3 Which Technique Is Better?

The simple answer to this question is "it depends on what you want." The HPLC-based measurement of 8-OHdG is a highly sensitive method, largely because of the use of electrochemical detection, introduced by Floyd et al. (1986). Floyd et al. (1986) quote a detection limit of 20 fmol, or one 8-OHdG per 10^6 nucleosides. Shigenaga et al. (1990) quote 5–50 fmol on 40–100 μg samples of DNA.

How much 8-OHdG does the technique measure in cellular DNA? Floyd (1990) quotes background levels in cells or tissues (not deliberately subjected to oxidative stress) as 0.5–2.0 8-OHdG residues per 10^5 guanosines, or between 1 and 5 8-OHdG residues per 10^6 DNA bases (assuming that DNA has, on average, 25% guanine bases). Ames (1989) gives similar figures (one 8-OHdG per 130 000 bases in rat liver nuclear DNA, or 8 per 10^6 bases, but 1/8000 in mitochondrial DNA, or 125 per 10^6 bases). Most other scientists obtain broadly comparable results

Table 2. "Baseline" levels of 8-hydroxydeoxyguanosine in DNA as determined by high performance liquid chromatography (calculation assumes that guanine is, on average, 25% of the DNA bases)

Source of DNA	Content of 8-OHdG		Reference
	Per 10^5 guanines	Per 10^6 DNA bases	
Rat liver	1.21 ± 0.45	≈ 3	Conway et al. (1991)
(total DNA or	4.0 ± 1.3		
nuclear DNA)	3.1 ± 0.4	≈ 8	Denda et al. (1991)
	7.4 ± 0.9	≈ 19	Hinrichsen et al. (1990)
	1.3–3.3	3–8	Fraga et al. (1990)
		≈ 20	Heig et al. (1990)
Mouse liver	0.96 ± 0.37	2.4	Adachi et al. (1993)
(total DNA)	0.6–1.4	1.5–3.5	Kasai et al. (1986)
	≈ 20	≈ 50	Faux et al. (1992)
Hamster liver	≈ 7	≈ 17	Roy et al. (1991)
Rat liver	≈ 13	≈ 33	Richter et al. (1988)
(mitochondrial DNA)			
Human phagocytes	0.3 ± 0.008[a]	≈ 1	Kiyosawa et al. (1990)
	–	<10	Floyd et al. (1986)
Commercial calf	3–128[b]	8–320	Floyd et al. (1989)
thymus DNA	–	26.8 ± 12.6	Aiyar et al. (1990)
	22.2 ± 2	≈ 55	Kiyosawa et al. (1990)
	–	25	Mouret et al. (1991)
	–	70	Lu et al. (1991)
	6	15	Park and Floyd (1992)
Rat kidney DNA	2.4–4.7	≈ 6–12	Fraga et al. (1990)
	≈ 1.7	≈ 4	Sai et al. (1991)
	≈ 1	≈ 3	Umemura et al. (1991)
Hamster kidney DNA	≈ 3	≈ 7	Roy et al. (1991)
HeLa cell DNA	0.6–1.4	1.5–3.5	Kasai et al. (1986)
Salmonella typhimurium DNA	0.6–1.4	1.5–3.5	Kasai et al. (1986)

[a] From nonsmokers.
[b] Depending on batch of commercial DNA used.

Table 3. GC–MS-based measurement of 8-hydroxyguanine in DNA (after acidic hydrolysis of DNA, unless otherwise stated)

Source of DNA	8-OHG		Totally modified bases[a]	Reference
	nmol/mg DNA	Bases/10^6 DNA bases	(per 10^6 DNA bases)	
Commercial calf thymus DNA	0.5–1.0	159–318	≈ 640	Aruoma et al. (1989a,b, 1991)
Mouse liver[b] mitochondrial DNA	–	≈ 3500	–	Hayakawa et al. (1991)
Fish liver	0.13 ± 0.06	≈ 41	–	Malins and Haimanot (1990)
Chromatin isolated[c] from murine hybridoma cells	–	35–40	≈ 138	Dizdaroglu et al. (1991b)
Human breast tissue	0.13 ± 0.03	≈ 41	–	Malins and Haimanot (1991)
Human neutrophils	0.20–0.23	64–73	–	Malins and Haimanot (1991)
Human brain	–	14		Olinski et al. (1992)
Human lung	–	73–97		Olinski et al. (1992)

[a] Total of all the modified bases measured in the DNA.
[b] GC/MS analysis after enzymic hydrolysis of DNA.
[c] Freshly isolated chromatin. Chromatin after extensive dialysis gives higher figures.

(some published data are summarized in Table 2). The range is between 3 and 50 8-OHdG per 10^6 DNA bases. Some part of the variation might be accounted for by an effect of age (Fraga et al. 1990) in some rat tissues. By contrast, commercial calf thymus DNA (frequently used in studies in vitro) contains much more 8-OHdG than freshly isolated DNA (Table 2). Floyd et al. (1989) quote a range of 8–320 8-OHdG per 10^6 bases, depending on the batch of commercial DNA used.

The GC–MS–SIM technique also provides high sensitivity and selectivity and can be applied directly to chromatin (Dizdaroglu 1991; Dizdaroglu et al. 1991a). The highest sensitivity for a compound under analysis (generally about 5 fmol per compound for products of DNA base modification) is achieved by monitoring the most abundant characteristic ion in its mass spectrum. For initial processing of DNA samples, 50–100 mg DNA has usually been used but smaller amounts are feasible since usually only 0.1–0.4 mg hydrolyzed and derivated DNA is injected onto the GC column for actual analysis.

The lowest background level of a base modification in DNA or in chromatin measurable by currently used GC–MS–SIM techniques corresponds to 1 modified base in approximately 10^6 bases (Dizdaroglu 1991). This makes the technique broadly comparable in sensitivity to measurement of 8-OHdG by HPLC with electrochemical detection. The exact sensitivity achieved is affected by the GC–MS instrument used and the type of column. In most cases, it is not the absolute sensitivity of the technique that matters, but the "background" levels of base modification in untreated DNA, or DNA from "unstressed" cells or tissues (e.g., Lutgerink et al. 1992). Comparison of Tables 2 and 3 shows that GC–MS–SIM is sensitive enough to detect the levels of 8-OHdG that have been recorded by HPLC analysis. GC–MS has been used to show that human cancers appear to contain elevated levels of DNA base damage products (Malins and Haimanot 1991; Olinski et al. 1992), suggestive of increased oxidative damage and/or decreased repair of such damage in vivo.

Commercial calf thymus DNA usually contain 0.5–1.0 nmol of 8-OHG per mg, as measured by GC–MS after formic acid hydrolysis and trimethylsilylation. This corresponds to 159–318 bases/10^6 bases, within the range of HPLC-determined values stated by Floyd et al. (1989) (see Table 2). Values up to 570 8-OHdG per 10^5 guanines were also obtained in commercial calf thymus DNA using a ^{32}P-postlabeling technique (Lutgerink et al. 1992). To date, fewer studies on DNA freshly isolated from cells and tissues have been done with GC–MS–SIM than with HPLC, but the figures available show around 40 8-OHG per 10^6 DNA bases (Table 3), generally higher than the figures recorded by HPLC (Table 2).

3.5.4 Explanations of the Discrepancy

Modifications of DNA bases affect cell metabolism and may be related to carcinogenesis (Cerutti et al. 1989; Lindahl 1993; Olinski et al. 1992; Ames 1989; Breimer 1990, 1991), so it is important to understand this apparent discrepancy. What are the possible explanations? First, HPLC may underestimate the real amount of 8-OHdG in DNA. Second, GC–MS may overestimate it, e.g., if the derivatization and hydrolysis procedures somehow artifactually cause formation of 8-OHG. Third, both errors could occur. Before discussing these potential explanations in detail, it is worth pointing out that isolating of DNA from cells may introduce some oxidative modification, particularly if phenol-based methods are used (Claycamp 1992), since oxidizing phenols produce a wide range of reactive radicals (Halliwell and Gutteridge 1989). This is one of the reasons why extraction of chromatin for analysis may be preferable (Dizdaroglu 1991; Dizdaroglu et al. 1991a). This technique is milder than those used for DNA extraction and may minimize the loss of extensively fragmented DNA and of DNA that has become cross-linked to protein as a result of oxidative damage.

3.5.5 How Could HPLC Underestimate 8-OHdG?

Once "up and running," the HPLC technique is fairly reliable and does not treat the DNA as "roughly" as do the preparation procedures that are used for GC–MS. However, HPLC analysis requires complete extraction of DNA from the cell and complete enzymic hydrolysis before quantitative measurement of 8-OHdG can be achieved. First, extraction of DNA that has undergone extensive oxidative modification and fragmentation may be impaired, because of the easy loss of small DNA fragments and of cross-linking of the DNA bases to amino acid residues in nuclear proteins (Dizdaroglu 1991). Hence this HPLC method may underestimate DNA damage in cells that have been subjected to intense oxidative stress.

Second, the efficiency of exonucleases and endonucleases in hydrolyzing DNA is greatly affected by modification of the bases (Breimer 1991; Dizdaroglu et al. 1978). For example, Maccubbin et al. (1991) reported that the presence of 8-OHG severely inhibits digestion of dinucleotides by phosphodiesterase. Thus it is not always certain that modified bases are completely hydrolyzed from DNA, especially when published hydrolysis techniques are transplanted from one laboratory to another and not revalidated.

Third, the HPLC technique as usually used measures 8-OHdG, and not 8-OHG, which is separated on the columns used. Frenkel et al. (1991) showed that acid pH (frequently used for nuclease P_1 digestion) can promote hydrolysis of 8-OHdG to 8-OHG, causing a loss of HPLC-detectable material. Extensive free radical damage might also lead to release of modified guanine from the DNA backbone to leave an abasic site. Frenkel et al. (1991) considered these artifacts and, using a more complex HPLC technique than is commonly employed, they found that DNA extracted from murine epidermal cells contains baseline 8-OHdG levels of at least 30 per 10^6 DNA bases, closer to the values measured by GC-MS (Table 3) than to those measured by HPLC.

3.5.6 Artifacts in GC–MS Techniques?

The GC–MS techniques give much more chemical information than does measurement of only 8-OHdG. However, the hydrolysis and derivatization procedures are lengthy and tedious and may destroy some modified bases, e.g., hydroxymethyluracil (Djuric et al. 1991). It has also been speculated that they might create artifacts, for example, if the amounts of modified bases increase during the preparation procedures. It has been shown that formic acid hydrolysis does not create additional 8-OHG in DNA (Halliwell and Dizdaroglu 1992), but the question is currently open as to whether the derivatization procedures might do so.

3.6 Conclusion

We do not know the "baseline" level of products of DNA damage by oxygen-derived species in vivo, since different measurement techniques give different results. Greater attention must be given to resolving these methodological questions and validating the assays being used before the widespread adoption of such methods as a true index of oxidative DNA damage.

Acknowledgments. I thank the Medical Research Council, Arthritis and Rheumatism Council and Cancer Research Campaign for research support.

References

Adachi S, Takemoto K, Hirosue T, Hosogai Y (1993) Spontaneous and 2-nitropropane induced levels of 8-hydroxy-2'-deoxyguanosine in liver DNA of rats fed iron-deficient or manganese- and copper-deficient diets. Carcinogenesis 14:265–268

Agarwal RP, Parks Jr RE (1978) Adenosine deaminase from human erythrocytes. Meth Enzymol L1:502–507

Aida M, Nishimura S (1987) An ab initio molecular orbital study on the characteristics of 8-hydroxyguanine. Mut Res 192:83–89

Aiyar J, Berkovits HJ, Floyd RA, Wetterhahn KE (1990) Reaction of chromium (VI) with hydrogen peroxide in the presence of glutathione: reactive intermediates and resulting DNA damage. Chem Res Tox 3:595–603

Alam S, Ali S, Ali R (1993) The effect of hydroxyl radical on the antigenicity of native DNA. FEBS Lett 319:66–70

Ames BN (1989) Endogenous oxidative DNA damage, aging and cancer. Free Rad Res Comms 7:121–128

Aruoma OI, Halliwell B, Dizdaroglu M (1989a) Iron ion-dependent modification of bases in DNA by the superoxide radical-generating system hypoxanthine/oxidase. J Biol Chem 264:13024–13028

Aruoma OI, Halliwell B, Gajewski E, Dizdaroglu M (1989b) Damage to the bases in DNA induced by hydrogen peroxide and ferric ion chelates. J Biol Chem 264:20509–20512

Aruoma OI, Halliwell B, Gajewski E, Dizdaroglu M (1991) Copper-ion-dependent damage to the bases in DNA in the presence of hydrogen peroxide. Biochem J 273:601–604

Beckman JS, Beckman TW, Chen J, Marshall PA, Freeman BA (1990) Apparent hydroxyl radical production by peroxynitrite: implications for endothelial injury from nitric oxide and superoxide. Proc Natl Acad Sci USA 87:1620–1624

Berger M, Anselmino C, Mouret JF, Cadet J (1990) High performance liquid chromatography-electrochemical assay for monitoring the formation of 8-oxo-7,8-dihydro-adenine and its related 2'-deoxyribonucleoside. J Liquid Chromatog 13:929–940

Birnboim HC (1988) A superoxide anion induced DNA strand-break metabolic pathway in human leukocytes: effect of vanadium. Biochem Cell Biol 66:374–381

Blount S, Griffiths HR, Staines NA, Lunec J (1992) Probing molecular changes induced in DNA by reactive oxygen species with monoclonal antibodies. Immunol Lett 34:115–126

Breimer LH (1988) Ionizing radiation-induced mutation. Br J Cancer 57:6–18

Breimer LH (1990) Molecular mechanisms of oxygen radical carcinogenesis and mutagenesis. The role of DNA base damage. Mol Carcinog 3:188–197

Breimer LH (1991) Repair of DNA damage induced by reactive oxygen species. Free Rad Res Comms 14:159–171

Burdon RH, Rice-Evans C (1989) Free radicals and the regulation of mammalian cell proliferation. Free Rad Res Comms 6:346–348

Canonero R, Martelli A, Marinari UR, Brambilla G (1990) Mutation induction in Chinese hamster lung V79 cells by five alk-2-enals produced by lipid peroxidation. Mutat Res 244:153–156

Cantoni O, Sestili P, Cattabeni F, Bellomo G, Pou S, Cohen M, Cerutti P (1989) Calcium chelator quin 2 prevents hydrogen-peroxide-induced DNA breakage and cytotoxicity. Eur J Biochem 181:209–212

Carmichael PL, She MN, Phillips DH (1992) Detection and characterization by ^{32}P-postlabelling of DNA adducts induced by a Fenton-type oxygen radical-generating system. Carcinogenesis 13:1127–1135

Cerutti P, Larsson R, Krupitza G, Muehlmatter D, Crawford D, Amstad P (1989) Pathophysiological mechanisms of active oxygen. Mutat Res 214:81–88

Cheng KC, Cahill DS, Kasai H, Nishimura S, Loeb LA (1992) 8-Hydroxy-guanine, an abundant form of oxidative DNA damage, causes G → T and A → C substitutions. J Biol Chem 267:166–172

Chung FL, Xu Y (1992) Increased 8-oxodeoxyguanosine levels in lung DNA of A/J mice and F344 rats treated with the tobacco-specific nitrosamine 4-(methyl-nitrosamine)-1-(3-pyridyl)-1-butanone. Carcinogenesis 13:1269–1272

Claycamp HG (1992) Phenol sensitization of DNA to subsequent oxidative damage in 8-hydroxyguanine assays. Carcinogenesis 13:1289–1292

Conway CC, Nie G, Huesain NS, Fiala ES (1991) Comparison of oxidative damage to rat liver DNA and RNA by primary nitroalkanes, secondary nitroalkanes, cyclopentanone oxime and related compounds. Cancer Res 51:3143–3147

Denda A, Sai K, Tang Q, Tsujuchi T, Tsutsumi M, Amanuwa T, Murata Y, Nakoe D, Maruyama H, Kurokawa Y, Konishi Y (1991) Induction of 8-hydroxydeoxyguanosine but not initiation of carcinogenesis by redox enzyme modulations with or without menadione in rat liver. Carcinogenesis 12:719–726

Devasagayam TPA, Steenken S, Obendorf MSW, Schultz WA, Sies H (1991) Formation of 8-hydroxy(deoxy)guanosine and generation of strand breaks at guanine residues in DNA by singlet oxygen. Biochemistry 30:6283–6289

Di Guiseppi G, Fridovich I (1984) The toxicology of molecular oxygen. CRC Crit Rev Toxicol 12:315–342

Dijkwel PA, Wenink PW (1986) Structural integrity of the nuclear matrix; differential effects of thiol agents and metal chelators. J Cell Sci 84:53–67

Dizdaroglu M (1991) Chemical determination of free radical-induced damage to DNA. Free Rad Biol Med 10:225–242

Dizdaroglu M (1993) Quantitative determination of oxidative base damage in DNA by stable isotope-dilution mass spectrometry. FEBS Lett 315:1–6

Dizdaroglu M, Hermes W, Schulte-Frohlinde D, von Sonntag C (1978) Enzymatic digestion of DNA γ-irradiated in aqueous solution. Separation of the digests by ion-exchange chromatography. Int J Radiat Biol 33:563–569

Dizdaroglu M, Rao G, Halliwell B, Gajewski E (1991a) Damage to the DNA bases in mammalian chromatin by hydrogen peroxide in the presence of ferric and cupric ions. Arch Biochem Biophys 285:317–324

Dizdaroglu M, Nackerdien Z, Chao BC, Gajewski E, Rao G (1991b) Chemical nature of in vivo DNA base damage in hydrogen peroxide-treated mammalian cells. Arch Biochem Biophys 285:388–390

Djuric A, Luongo DA, Harper DA (1991) Quantitation of 5-(hydroxymethyl) uracil in DNA by gas chromatography with mass spectral detection. Chem Res Tox 4:687–691

Epe B, Mutzel R, Adam W (1988) DNA damage by oxygen radicals and excited state species: a comparative study using enzymatic probes in vitro. Chem-Biol Interac 67:149–165

Epe B, Pflaum M, Haring M, Hegler J, Rudiger H (1993) Use of repair endonucleases to characterize DNA damage induced by reactive oxygen species in cellular systems. Toxicol Lett 67:57–72

Essigmann JM, Wood ML (1993) The relationship between the chemical structures and mutagenic specificities of the DNA lesions formed by chemical and physical mutagens. Toxicol Lett 67:29–39

Evans J, Maccabee M, Hatahet Z, Courcelle J, Bockrath R, Ide H, Wallace S (1993) Thymine ring saturation and fragmentation products: lesion bypass, misinsertion and implications for mutagenesis. Mut Res 299:147–156

Farber JL (1990) The role of calcium in lethal cell injury. Chem Res Toxicol 3:503–508

Faux SP, Francis JE, Smith AG, Chipman JK (1992) Induction of 8-hydroxydeoxyguanosine in Ah-responsive mouse liver by iron and Aroclor 1254. Carcinogenesis 13:247–250

Feig DI, Loeb LA (1993) Mechanisms of mutation by oxidative DNA damage: reduced fidelity of mammalian DNA polymerase β. Biochemistry 32:4466–4473

Floyd RA (1990) The role of 8-hydroxyguanine in carcinogenesis. Carcinogenesis 11:1447–1450

Floyd RA, Watson JJ, Wong PK, Altmiller DH, Rickard RC (1986) Hydroxylfree radical adduct of deoxyguanosine: sensitive detection and mechanisms of formation. Free Rad Res Comms 1:163–172

Floyd RA, West MS, Eneff KL, Hogsett WE, Tingey DT (1988) Hydroxyl free radical mediated formation of 8-hydroxyguanine in isolated DNA. Arch Biochem Biophys 262:266–272

Floyd RA, West MS, Eneff KL, Schneider JE (1989) Methylene blue plus light mediates 8-hydroxyguanine formation in DNA. Arch Biochem Biophys 273:106–111

Fraga CC, Shigenaga MK, Park JW, Degan P, Ames BN (1990) Oxidative damage to DNA during aging: 8-hydroxy-2'-deoxyguanosine in rat organ DNA and urine. Proc Natl Acad Sci USA 87:4533–4537

Fraga CG, Motchnik PA, Shigenaga MK, Helbock HJ, Jacob RA, Ames BN (1991) Ascorbic acid protects against endogenous oxidative DNA damage in human sperm. Proc Natl Acad Sci USA 88:11006–11033

Frebourg T, Friend SH (1992) Cancer risks from germline P53 mutations. J Clin Invest 90:1637–1641

Frenkel K (1992) Carcinogen-mediated oxidant formation and oxidative DNA damage. Pharmac Ther 53:127–166

Frenkel K, Zhong Z, Wei H, Karkoszka J, Patel U, Rashid K, Georgescu M, Solomon JJ (1991) Quantitative high-performance liquid chromatography analysis of DNA oxidized in vitro and in vivo. Anal Biochem 196:126–136

Gajewski E, Rao G, Nackerdien Z, Dizdaroglu M (1990) Modification of DNA bases in mammalian chromatin by radiation-generated free radicals. Biochemistry 29:7876–7882

Geborek P, Mansson B, Hellmer G, Saxne T (1992) Cytidine deaminase and lactoferrin in inflammatory synovial fluids – indicators of local polymorphonuclear cell function. Br J Rheumatol 31:235–240

Geierstanger BH, Kagawa TF, Chen SL, Quigley GJ, Ho PS (1991) Base-specific binding of copper (II) to Z-DNA. J Biol Chem 266:20185–20191

Glantz MD, Lewis AS (1978) Guanine deaminase from rabbit liver. Meth Enzymol L1:512–517

Goligorsky MS, Morgan MA, Suh H, Safirstein R, Johnson R (1992) Mild oxidative stress: cellular mode of mitogenic effect. Renal Failure 14:385–389

Gutteridge JMC, Halliwell B (1990) The measurement and mechanism of lipid peroxidation in biological systems. Trends Biochem Sci 15:129–135

Hadley M, Draper HH (1990) Isolation of a guanine-malondialdehyde adduct from rat and human urine. Lipids 25:82–85

Halliwell B (1987) Oxidants and human disease; some new concepts. FASEB J 1:358–362

Halliwell B, Aruoma OI (1991) DNA damage by oxygen-derived species. Its mechanism and measurement in mammalian systems. FEBS Lett 281:9–19

Halliwell B, Dizdaroglu M (1992) The measurement of oxidative damage to DNA by HPLC and GC/MS techniques. Free Rad Res Comms 16:75–87

Halliwell B, Gutteridge JMC (1984) Lipid peroxidation, oxygen radicals, cell damage and antioxidant therapy. Lancet 1:1396–1398

Halliwell B, Gutteridge JMC (1989) Free radicals in biology and medicine, 2nd edn. Clarendon, Oxford

Halliwell B, Gutteridge JMC (1990a) Role of free radicals and catalytic metal ions in human disease. Methods Enzymol 186:1–185

Halliwell B, Gutteridge JMC (1990b) The antioxidants of human extracellular fluids. Arch Biochem Biophys 280:1–8

Hayakawa M, Ogawa T, Sugiyama S, Tanaka M, Ozawa T (1991) Massive conversion of guanine to 8-hydroxyguanosine in mouse liver mitochondrial DNA by administration of azidothymidine. Biochem Biophys Res Commun 176:87–93

Hayakawa M, Hattori K, Sugiyama S, Ozawa T (1992) Age-associated oxygen damage and mutations in mitochondrial DNA in human hearts. Biochem Biophys Res Commun 189:979–985

Heig ME, Ulrich D, Sagelsdorff P, Richter C, Lutz WK (1990) No measurable increase in thymidine glycol or 8-hydroxydeoxyguanosine in liver DNA of rats treated with nafenopin or choline-devoid low-methionine diet. Mut Res 238:325–329

Hinrichsen LI, Floyd RA, Sudilovsky O (1990) Is 8-hydroxydeoxyguanosine a mediator of carcinogenesis by a choline-devoid diet in the rat liver? Carcinogenesis 11:1879–1881

Imlay JA, Linn S (1988) DNA damage and oxygen radical toxicity. Science 240:1302–1309

Kasai H, Nishimura S (1984) Hydroxylation of deoxyguanosine at the C-8 position by ascorbic acid and other reducing agents. Nucleic Acids Res 12:2137–2145

Kasai H, Tanooka H, Nishimura S (1984) Formation of 8-hydroxyguanine residues in DNA by X-irradiation. Gann 75:1037–1039

Kasai H, Crain PF, Kuchino Y, Nishimura S, Ootsuyama A, Tanooka H (1986) Formation of 8-hydroxyguanine moiety in cellular DNA by agents producing oxygen radicals and evidence for its repair. Carcinogenesis 7:1849–1851

Kasprzak KS (1991) The role of oxidative damage in metal carcinogenicity. Chem Res Tox 4:604–615

Kasprzak KS, Diwan BA, Rice JM, Misra M, Riggs CW, Olinski R, Dizdaroglu M (1992) Nickel(II)-mediated oxidative DNA base damage in renal and hepatic chromatin of pregnant rats and their fetuses. Possible relevance to carcinogenesis. Chem Res Tox 5:810–815

Kass GE, Duddy SK, Ossenius S (1989) Activation of hepatroyte protein kinase C by redox-cycling quinones. Biochem J 260:499–507

Kelly JD, Orner GA, Hendricks JD, Williams DE (1992) Dietary hydrogen peroxide enchances hepato-carcinogenesis in trout: correlation with 8-hydroxy-2'-deoxyguanosine levels in liver DNA. Carcinogenesis 13:1639–1642

Kiyosawa T, Suko M, Okudaira H, Murata K, Miyamoto T, Chung MH, Kasai H Nishimura S (1990) Cigarette smoke induces formation of 8-hydroxydeoxyguanosine, one of the oxidative DNA damages in human peripheral leukocytes. Free Rad Res Comms 11:23–27

Kojima T, Nishina T, Kitamura M, Kamatani N, Nishioka K (1989) Reversedphase high-performance liquid-chromatography of 2,8-dihydroxyadenine in serum and urine with electrochemical detection. Clin Chim Acta 181:109–114

Kyle ME, Nakae D, Sakaida I, Miccadei S, Farber JL (1988) Endocytosis of superoxide dismutase is required in order for the enzyme to protect hepatocytes from the cytotoxicity of hydrogen peroxide. J Biol Chem 263:3784–3789

Larrick JW, Wright SC (1990) Cytotoxic mechanism of tumor necrosis factor alpha. FASEB J 4:3215–3223

Larsson R, Cerutti P (1989) Translocation and enhancement of phosphotransferase activity of protein kinase C following exposure of mouse epidermal cells to oxidants. Cancer Res 49:5627–5632

Leanderson P, Tagesson C (1989) Cigarette smoke potentiates the DNA-damaging effect of manmade mineral fibres. Am J Indust Med 16:697–706

Leanderson P, Tegesson C (1992) Cigarette smoke-induced DNA damage in cultured human lung cells: role of hydroxyl radicals and endonuclease activation. Chem Biol Interac 81:197–208

Lewis CD, Laemmli UK (1982) Higher order metaphase chromosome structure: evidence for metalloprotein interactions. Cell 29:171–181

Lindahl T (1993) Instability and decay of the primary structure of DNA. Nature 362:709–715

Loft S, Vistisen K, Ewertz M, Tjonneland A, Overvad K, Poulsen HE (1992) Oxidative DNA damage estimated by 8-hydroxydeoxyguanosine excretion in humans: influence of smoking, gender and body mass index. Carcinogenesis 13:2441–2447

Lu LJW, Tasake F, Hokanson JA, Kohda K (1991) Detection of 8-hydroxy-2'deoxyguanosine in deoxyribonucleic acid by the ^{32}P-postlabelling method. Chem Pharm Bull 39:1880–1882

Lutgerink JT, de Graaf E, Hoebee B, Staenuitez HFC, Westra JG, Kriek E (1992) Detection of 8-hydroxyguanine in small amounts of DNA by ^{32}P postlabelling. Anal Biochem 201:127–133

Maccubbin A, Evans M, Paul CR, Budzinski EL, Przybyszewski J, Box HC (1991) Enzymatic excision of radiation-induced lesions from DNA model compounds. Radiat Res 126:21–26

Malins DC, Haimanot R (1990) 4,6,-Diamino-5-formamido-pyrimidine, 8-hydroxyguanine and 8-hydroxyadenine in DNA from neoplastic liver of English sole exposed to carcinogens. Biochem Biophys Res Commun 173:614–619

Malins DC, Haimanot R (1991) Major alterations in the nucleotide structure of DNA in cancer of the female breast. Cancer Res 51:5430–5432

McBride TJ, Preston BD, Loeb LA (1991) Mutagenic spectrum resulting from DNA damage by oxygen radicals. Biochemistry 30:207–213

Meier B, Radeke H, Selle S, Raspe HH, Sies H, Resch K, Habermehl GG (1990) Human fibroblasts release reactive oxygen species in response to treatment with synovial fluids from patients suffering from arthritis. Free Rad Res Comms 8:149–160

Mello-Filho AC, Hoffman RE, Meneghini R (1984) Cell killing and DNA damage by hydrogen peroxide are mediated by intracellular iron. Biochem J 218:273–276

Mo JY, Maki H, Sekiguchi M (1992) Hydrolytic elimination of a mutagenic nucleotide, 8-oxodGTP, by human 18-kilodalton protein: sanitization of nucleotide pool. Proc Natl Acad Sci USA 89:11021–11025

Moraes EC, Keyse SM, Tyrrell RM (1990) Mutagenesis by hydrogen peroxide treatment of mammalian cells: a molecular analysis. Carcinogenesis 11:283–293

Mouret JF, Polverelli M, Sarrazini F, Cadet J (1991) Ionic and radical oxidations of DNA by hydrogen peroxide. Chem-Biol Interac 77:187–201

Murrell GAC, Francis MJO, Bromley L (1990) Modulation of fibroblast proliferation by oxygen free radicals. Biochem J 265:659–665

Nackerdien Z, Rao G, Cacciuttolo MA, Gajewski E, Dizdaroglu M (1991) Chemical nature of DNA-protein cross-links produced in mammalian chromatin by hydrogen peroxide in the presence of iron or copper ions. Biochemistry 30:4872–4879

Nackerdien Z, Olinski R, Dizdaroglu M (1992) DNA base damage in chromatin of γ-irradiated human cells. Free Rad Res Comms 16:259–273

Nakae D, Yoshiji H, Amanuma T, Kinugasa T, Farber JL, Konishi Y (1990) Endocytosis-independent uptake of liposome-encapsulated superoxide dismutase prevents the killing of cultured hepatocytes by tert-butyl hydroperoxide. Arch Biochem Biophys 279:315–319

Nassi-Calo L, Mello-Filho AC, Meneghini R (1989) O-phenanthroline protects mammalian cells from hydrogen peroxide-induced gene mutation and morphological transformation. Carcinogenesis 10:1055–1057

Oda Y, Uesugi S, Ikehara M, Nishimura S, Kawase Y, Ishikawa H, Inoue H, Ohtsuka E (1991) NMR studies of a DNA containing 8-hydroxydeoxyguanosine. Nucleic Acids Res 19:1407–1412

Olinski R, Zastawny T, Budzbon J, Skokowski J, Zegarski W, Dizdaroglu M (1992) DNA base modifications in chromatin of human cancerous tissues. FEBS Lett 309:193–198

Oller AR, Thilly WG (1992) Mutational spectra in human β-cells. Spontaneous, oxygen and hydrogen peroxide-induced mutations at the hprt gene. J Mol Biol 228:813–826

Orrenius S, McConkey DJ, Bellomo G, Nicotera P (1989) Role of Ca^{++} in toxic cell killing. Trends Pharmacol Sci 10:281–285

Park JW, Floyd RA (1992) Lipid peroxidation products mediate the formation of 8-hydroxydeoxyguanosine in DNA. Free Rad Biol Med 12:245–250

Prieto-Alamo MJ, Abril N, Pueyo C (1993) Mutagenesis in Escherichia coli K-12 mutants defective in superoxide dismutase or catalase. Carcinogenesis 14:237–244

Prutz WA, Butler J, Land EJ (1990) Interaction of copper (I) with nucleic acids. Int J Radiat Biol 58:215–234

Retel J, Hoebee B, Braun JEF, Lutgerink JT, van den Akker E, Wanamarta AH, Joenje H, Lafleur MJM (1993) Mutational specificity of oxidative DNA damage. Mut Res 299:165–182

Richter C (1988) Do mitochondrial DNA fragments promote cancer and aging? FEBS Lett 241:1–5

Richter C, Park JW, Ames BN (1988) Normal oxidative damage to mitochondrial and nuclear DNA is extensive. Proc Natl Acad Sci USA 85:6465–6467

Rowley DA, Halliwell B (1983) DNA damage by superoxide-generating systems in relation to the mechanism of action of the anti-tumor antibiotic adriamycin. Biochim Biophys Acta 761:86–93

Roy D, Floyd RA, Liehr JG (1991) Elevated 8-hydroxy-deoxyguanosine levels in DNA of diethylstilbestrol-treated Syrian hamsters; covalent DNA damage by free radicals generated by redox cycling of diethylstilbestrol. Cancer Res 51:3882–3885

Sai K, Takagi A, Umemura T, Hasegawa R, Kurokawa Y (1991) Relation of 8-hydroxydeoxyguanosine formation in rat kidney to lipid peroxidation, glutathione level and relative organ weight after a single administration of potassium bromate. Japan J Cancer Res 82:165–169

Shibutani S, Takeshita M, Grollman AP (1991) Insertion of specific bases during DNA synthesis past the oxidation-damaged base 8-oxodG. Nature 349:431–434

Shigenaga MK, Park JW, Cundy KC, Gimeno CJ, Ames BN (1990) In vivo oxidative DNA damage: measurement of 8-hydroxy-2'-deoxyguanosine in DNA and urine by high-performance liquid chromatography with electrochemical detection. Meth Enzymol 186:521–530

Sies H ed (1991) Oxidative stress, oxidants and antioxidants. Academic, New York

Sies H (1993) Damage to plasmid DNA by singlet oxygen and its protection. Mut Res 299:183–191

Sodum RS, Chung FL (1988) INZ-ethenodeoxyguanosine as a potential marker for DNA adduct formation by trans-4-hydroxy-2-nonenal. Cancer Res 48:320–323

Steenken S (1989) Purine bases, nucleosides and nucleotides: aqueous solution redox chemistry and transformation reactions of their radical cations and e⁻ and OH adducts. Chem Rev 89:503–520

Stillwell WG, Xu HX, Adkins JA, Wishnok JS, Tannenbaum SR (1989) Analysis of methylated and oxidized purines in urine by capillary gas chromatography-mass spectrometry. Chem Res Tox 2:94–99

Stone K, Ksebati M, Marnett LJ (1990a) Identification of the adducts formed by reaction of malondialdehyde with adenosine. Chem Res Tox 3:33–38

Stone K, Uzieblo A, Marnett LJ (1990b) Studies of the reaction of malondialdehyde with cytosine nucleosides. Chem Res Tox 3:467–472

Tchou J, Grollman AP (1993) Repair of DNA containing the oxidatively-damaged base, 8-oxoguanine. Mut Res 299:277–287

Totter JR (1980) Spontaneous cancer and its possible relationship to oxygen metabolism. Proc Natl Acad Sci USA 77:1763–1767

Touati D (1989) The molecular genetics of superoxide dismutase in E. coli. Free Rad Res Comms 8:1–8

Trush MA, Kensler TW (1991) An overview of the relationship between oxidative stress and chemical carcinogenesis. Free Rad Biol Med 10:201–209

Ueda N, Shah SV (1992) Endonuclease-induced DNA damage and cell death in oxidant injury to renal tubular epithelial cells. J Clin Invest 90:2593–2597

Umemura T, Sai K, Takagi A, Hasegawa R, Kurokawa Y (1991) The effects of exogenous glutathione and cysteine on oxidative stress induced by ferric nitrilotriacetate. Cancer Lett 58:49–56

von Sonntag C (1987) The chemical basis of radiation biology. Taylor and Francis, London

Wagner JR, Hu CC, Ames BN (1992) Endogenous oxidative damage of deoxycytidine in DNA. Proc Natl Acad Sci USA 89:3380–3384

Weiss SJ (1989) Tissue destruction by neutrophils. N Engl J Med 320:365–376

Weitberg AB, Corvese D (1990) Translocation of chromosomes 16 and 18 in oxygen radical-transformed human lung fibroblasts. Biochem Biophys Res Commun 169:70–74

Weitzman SA, Gordon LI (1990) Inflammation and cancer: role of phagocyte-generated oxidants in carcinogenesis. Blood 76:655–663

Weitzman SA, Weitberg AB, Clark EP, Clark TP, Stossel TP (1985) Phagocytes as carcinogens: malignant transformation produced by human neutrophils. Science 227:1231–1233

Weitzman S, Schmeichel C, Turk P, Stevens C, Tolsma S, Bouck N (1988) Phagocyte-mediated carcinogenesis: DNA from phagocyte-transformed C3H 10T1/2 cells can transform NIH/3T3 cells. Ann NY Acad Sci 551:103–109

Wink DA, Kasprzak KS, Maragos CM, Elespura RK, Misra M, Dunams TM, Cebula TA, Koch WH, Andrews AW, Allen JS, Keefer LK (1991) DNA deaminating ability and genotoxicity of nitric oxide and its progenitors. Science 254:1001–1003

Wyllie AH (1980) Glucocorticoid-induced thymocyte apoptosis is associated with endogenous endonuclease activation. Nature 284:555–556

Zimmerman R, Cerutti P (1984) Active oxygen acts as a promoter of carcinogenesis in C3H/10T1/2/C18 fibroblasts. Proc Natl Acad Sci USA 81:2085–2087

Zimmerman RJ, Chan A, Leadon SA (1989) Oxidative damage in murine tumor cells treated in vitro by recombinant human tumor necrosis factor. Cancer Res 49:1644–1648

4 $\alpha_{2\mu}$-Globulin Mediated Male Rat Kidney Carcinogenesis

J. A. Swenberg

4.1 Introduction

$\alpha_{2\mu}$-Globulin ($\alpha_{2\mu}$) nephropathy is a well-characterized disease that only occurs in male rats. Female rats and either sex of mice, guinea pigs, hamsters, dogs, and monkeys do not develop this disease when exposed to agents which cause this syndrome in male rats (Borghoff et al. 1990; Alden 1986; Swenberg et al. 1989; Halder et al. 1985). Likewise, there are no data suggesting that these agents cause protein droplet accumulation in the kidneys of humans. Research from several laboratories has elucidated the mechanism responsible for this remarkable sex- and species-specific disease. This understanding of mechanisms led the EPA (1991) to conclude that renal tumors mediated through $\alpha_{2\mu}$ should not be used in human risk assessment.

The acute disease is characterized by the accumulation of $\alpha_{2\mu}$ in lysosomes of the P_2 segment of the nephron, single cell necrosis, formation of granular casts at the junction of P_3 and the thin loop of Henle, compensatory cell proliferation, and the presence of regenerative tubules (Borghoff et al. 1990; Alden 1986; Swenberg et al. 1989; Halder et al. 1985). Following chronic exposure to agents causing $\alpha_{2\mu}$ nephropathy, there is an exacerbation of these lesions, the formation of linear mineralization of the renal pelvis and the induction of a 0%–25% incidence of renal cell tumors. As mentioned above, neither female rats nor mice of either sex exhibit renal disease when similarly exposed.

$\alpha_{2\mu}$-Globulin is a low-molecular-weight protein of 18 700 daltons that is synthesized under androgenic control by the hepatic parenchymal cells of mature male rats, secreted into the blood, freely filtered through the glomerulus, partially resorbed by the P_2 segment of the nephron, and excreted into the urine in large amounts (3–19 mg/day) (Roy and Neuhaus 1966; Roy et al. 1966; Ekstrom and Hoekstra 1984; Vandoren et al. 1983). $\alpha_{2\mu}$ is also found in the urine of female rats, however, the concentrations are 107 to 680-fold lower than that found in male rats (Vandoren et al. 1983). This sex-specific difference is based on the lack of synthesis of the androgen-dependent hepatic form of $\alpha_{2\mu}$ in the female rat.

4.2 Mechanisms of $\alpha_{2\mu}$ Nephropathy and Carcinogenesis

The chemicals known to cause this disease and the extent of mechanistic data available for each are listed in Table 1. These chemicals or their metabolites can bind reversibly to $\alpha_{2\mu}$ (Lock et al. 1987; Charbonneau et al. 1989; Strasser et al. 1988; Lehman-McKeeman et al. 1989). Even though the agents fall into rather diverse chemical classes, molecular modeling studies have demonstrated strong structure activity relationships (Borghoff et al. 1991). Active chemicals fit deeply into a hydrophobic pocket of $\alpha_{2\mu}$. When hydrogen bonding between the chemical and protein can occur, the digestibility of $\alpha_{2\mu}$ by proteases is inhibited, leading to accumulation of the male rat-specific protein in lysosomes of the P_2 segment of the nephron (Lehman-McKeeman et al. 1990). Hyaline droplets can also be produced by injecting female rats i.p. with $\alpha_{2\mu}$ (Ridder et al. 1990). This accumulation of $\alpha_{2\mu}$ is cytotoxic and results in single cell necrosis. The exfoliated renal epithelium, which represents the nidus for granular cast formation, is restored by compensatory cell proliferation.

This increase in cell proliferation is localized in the P_2 segment of the nephron and to a much lesser extent in the P_3 segment (Short et al. 1986, 1987, 1989a). Increased cell proliferation can be readily demonstrated using pulse (Short et al. 1986) or continuous (Short et al. 1987, 1989a; Dietrich and Swenberg 1991b) administration of [^{3}H]thymidine or bromodeoxyuridine, can be detected as early as 3 days after exposure to $\alpha_{2\mu}$-inducing agents, and has been demonstrated to remain elevated through at least 50 weeks of exposure (Short et al. 1989a). The extent of this increase in proliferation is dose-related, with maximum tolerated doses (MTDs) resulting in 5- to 12-fold greater numbers of labeled cells (Short et al. 1987, 1989a). Clear, no-effect doses can be demonstrated (Short et al. 1986, 1987, 1989a). Of great importance is the demonstration that female rats that have been identically exposed have no increase in cell proliferation (Short et al. 1989a). While this strongly suggested that the increase in cell proliferation requires the presence of large amounts of $\alpha_{2\mu}$, a recent study comparing cell proliferation in d-limonene exposed F344 versus NBR male rats has shown that the protein is absolutely required (Dietrich and Swenberg 1991b). The NBR rat is the only identified strain of rat that does not synthesize the androgen-dependent form of $\alpha_{2\mu}$ (Chatterjee et al. 1989). Whereas the F344 rats exposed to d-limonene exhibited a fivefold increase in cell proliferation after 5 or 30 weeks exposure to 150 mg/kg per day, NBR rats were unaffected (Dietrich and Swenberg 1991b). Both strains metabolized d-limonene to the 1,2-oxide, the nongenotoxic metabolite that reversibly binds to $\alpha_{2\mu}$ (Watabe et al. 1980; Dietrich and Swenberg 1991b). The increase in cell proliferation associated with $\alpha_{2\mu}$ nephropathy is reversible. Following exposures of up to 3 weeks to 2,2,4-trimethylpentane (TMP) or unleaded gasoline (UG), proliferation returns to control rates within 1 week after cessation of exposure (Short et al. 1989a). Longer term exposures result in a slower return to control rates. Morphologic evidence of regenerative tubules can still be identified 4 weeks after subchronic exposure ceases (Halder et al. 1984).

This sustained increase in cell proliferation is capable of promoting spontaneously initiated or chemically initiated cells of the proximal tubule to form preneoplastic and neoplastic lesions (Short et al. 1989b; Dietrich and Swenberg, 1991b). The promoting activity is totally dependent on the presence of $\alpha_{2\mu}$. When F344 rats were exposed to UG or TMP in an initiation-promotion study, concentration-related increases

Table 1. Data sets for chemicals causing α_{2u} globulin nephropathy[a]

Substance/chemical	Protein droplets	Increased α_{2u}	α_{2u} Binding	Cell proliferation	Initiation/promotion
d-Limonene	+	+	+	+	+
Unleaded gasoline	+	+	+	+	+
2,2,4-Trimethyl pentane	+	+	+	+	+
1,4-Dichlorobenzene	+	+	+	+	NR
Isophorone	+	+	+	+	NR
3,5,5-Trimethyl hexanoic acid derivatives	+	+	+	NR[b]	NR
Decalin	+	+	NR	+	NR
Tetrachloroethylene	+	+	NR	+	NR
Pentachloroethane	+	+	NR	+	NR
C10-C12 isoparaffinic solvent (saturated aliphatic hydrocarbons)	+	+	NR	NR	NR
Lindane	+	+	NR	NR	NR
BW540C	+	+	NR	NR	NR
BW58C	+	+	NR	NR	NR
Levamisole	+	+	NR	NR	NR
Gabapentin	+	+	NR	NR	NR
Tridecyl acetate	+	+	NR	NR	NR
Isoproplylcyclohexane	+	+	NR	NR	NR
JP-5 jet fuel (mixed distillate hydrocarbons)	+	NR	NR	NR	NR
JP-4 jet fuel (mixed distillate hydrocarbons)	+	NR	NR	NR	NR
Diesel, fuel marine	+	NR	NR	NR	NR
JP-10 synthetic jet fuel (exohexahydro-4,7-methanoindan)	+	NR	NR	NR	NR

Table 1. (cont.)

Substance/chemical	Protein droplets	Increased $\alpha_{2\mu}$	$\alpha_{2\mu}$ Binding	Cell proliferation	Initiation/promotion
RJ-5 synthetic jet fuel (hydrogenated dimers of norbornadiene)	+	NR	NR	NR	NR
JP-7 distillate jet fuel	+	NR	NR	NR	NR
JP-TS distillate jet fuel	+	NR	NR	NR	NR
Stoddard solvent	+	NR	NR	NR	NR
Tetralin	+	NR	NR	NR	NR
Hexachloroethane	+	NR	NR	NR	NR
Dimethyl methylphosphonate	+	NR	NR	NR	NR
Methyl isobutyl ketone	+	NR	NR	NR	NR
Methyl isoamyl ketone	+	NR	NR	NR	NR
Diisobutyl ketone	+	NR	NR	NR	NR
1,3,6-Tricyanohexane	+	NR	NR	NR	NR

NR, not reported
[a] Adapted from Appendix 1, US EPA (1991).
[b] Based on cell counts from urine.

in preneoplastic and neoplastic renal lesions were evident in males initiated with ethylhydroxyethylnitrosamine (EHEN) and promoted with UG or TMP (Short et al. 1989b). No increases occurred in females. The promoting activity paralleled increases in cell proliferation (Short et al. 1989a). When NBR rats were initiated with EHEN and promoted with d-limonene, no increase in atypical tubules, atypical hyperplasia, or renal adenomas occurred (Dietrich and Swenberg 1991b). In contrast, F344 rats promoted with d-limonene developed increased numbers of atypical tubules and atypical hyperplasia, while F344 rats initiated with EHEN and promoted with d-limonene developed increased numbers of atypical tubules, atypical hyperplasias and renal adenomas (Dietrich and Swenberg 1991b). Promotion of preneoplastic or neoplastic lesions only occurred in groups that also exhibited increased cell proliferation. An important observation from this study was the presence of occasional preneoplastic lesions in the kidneys of control rats of both strains, as these lesions are thought to represent early forms of spontaneous kidney tumors. The incidence of these lesions was increased by exposure to EHEN in both strains, and by exposure to d-limonene alone in F344 rats. These data strongly suggest that agents that cause $\alpha_{2\mu}$ nephropathy induce renal tumors in male rats through sustained increases in cell proliferation. The higher rate of cell proliferation decreases the amount of time available to repair DNA damage, increasing the probability of mutations leading to spontaneously initiated renal epithelial cells, and promoting clonal expansion of such cells, thereby increasing the probability of neoplasia.

Additional evidence for the sex and species specificity of this syndrome comes from studies on levamisole. Levamisole, a drug used as an antihelminthic in cancer chemotherapy and in the treatment of rheumatoid arthritis in humans, causes $\alpha_{2\mu}$-globulin nephropathy in male rats (Read et al. 1988). No increase in urinary N-acetyl β-glucosaminidase, an indicator of nephrotoxicity, was present in patients receiving 150 mg levamisole per day for 26 weeks (Dieppe et al. 1978). Levamisole has not yet been studied for carcinogenicity in animals or humans.

Several of the chemicals that induce $\alpha_{2\mu}$-related male rat kidney tumors also cause tumors at other sites. 1,4-Dichlorobenzene, pentachloroethane, tetrachloroethylene and unleaded gasoline also hepatocellular neoplasms in mice (National Toxicology Program 1983, 1986a,b, 1987; IARC 1987, 1989). Administration of tetrachloroethylene was also associated with an increased incidence of leukemia in rats (National Toxicology Program 1986). The mechanisms responsible for the induction of tumors at these sites are not known at this time.

Not all agents that induce $\alpha_{2\mu}$ nephropathy have resulted in an increased incidence of kidney tumors in male rats. In some cases, this has been due to an inadequate length of exposure. For example, a series of hydrocarbons was evaluated in rats exposed for 90 days and held for an additional 19 months (Bruner 1984, 1986). None of the hydrocarbons caused increases in renal tumors. When rats were exposed to hydrocarbons for 1 or more years, tumors were induced in the kidneys of male rats. The incidence of renal tumors induced by the $\alpha_{2\mu}$ mechanism is much lower than can be achieved by genotoxic renal carcinogens administered at their MTD. Furthermore, the extent of mechanistic research available for different agents varies markedly (see Table 1). Thus, while the weight of evidence available for this class of agents supporting this sex- and species-specific mechanism is very strong, determination of the relevance of this mechanism for a specific chemical's carcinogenic activity requires a case by case analysis.

4.3 Species Differences in Urinary Proteins

Having established that the presence of $\alpha_{2\mu}$ is mandatory for the formation of male rat kidney tumors following treatment with $\alpha_{2\mu}$ nephropathy-inducing agents, the question arises of whether extrapolation of such carcinogenicity data to other species, including humans, is warranted. Most compounds that are carcinogenic in animals are generally assumed to pose some risk to humans. In the case of $\alpha_{2\mu}$ nephropathy-inducing agents, the carcinogenic mechanism is clearly associated with the presence of a specific urinary protein ($\alpha_{2\mu}$) not found in humans or any other species. Several proteins sharing some amino acid sequence homology with $\alpha_{2\mu}$ have been identified in the serum and urine of various species, including humans (Pervaiz and Brew 1987; Akerstroem et al. 1987; Pevsner et al. 1988). The presence of these partially homologous proteins in humans raised concern as to the possible interaction of these low-molecular-weight proteins with $\alpha_{2\mu}$ nephropathy-inducing agents, thus questioning the male rat specificity of $\alpha_{2\mu}$ nephropathy. Assuming that a homologous protein reversibly binds the aforementioned chemicals, the protein needs to be excreted into the plasma in large amounts, freely filtered by the glomerulus, readily reabsorbed in the proximal tubules, and catabolized in the lysosomes of the proximal tubule epithelial cells at a slower rate than normal upon binding one of the chemicals in order to induce similar lesions to $\alpha_{2\mu}$ nephropathy. Based on estimated daily average urine production and body weights of rats and humans, Olson et al. (1990) showed that rats excrete approximately 90 times more total protein than humans. Of the total protein excreted, the predominant fraction in rats consisted of low-molecular-weight proteins (18 kDa), whereas a predominance of high-molecular-weight proteins (66 kDa) was found in humans. The small amount of low-molecular-weight protein excreted by male humans was identified as α_1-acid glycoprotein, α_1-microglobulin, myoglobin, and β_2-microglobulin. Of these four proteins only α_1-acid glycoprotein and α_1-microglobulin share amino acid sequence homology with $\alpha_{2\mu}$ (Pervaiz and Brew 1987; Akerstroem et al. 1987; Pevsner et al. 1988). α_1-Acid glycoprotein (AGP) and α_1-microglobulin (AMG) are synthesized in the liver of rats and humans (Ricca and Taylor 1981; Gross et al. 1987, 1988; Akerstroem and Landin 1985), have been purified from the urine of rats and humans, as well as the urine of rabbits and guinea pigs in the case of AMG (Akerstroem et al. 1987; Pevsner et al. 1988; Olson et al. 1990), and thus permit a direct comparison with $\alpha_{2\mu}$.

If AGP and AMG reversibly bind $\alpha_{2\mu}$ nephropathy-inducing chemicals and/or their metabolites with the same affinity as $\alpha_{2\mu}$, then one would expect that female rats, male NBR rats, rabbits, and guinea pigs would also develop renal disease following treatment with these chemicals. However, male NBR rats, female rats and guinea pigs do not accumulate protein in the renal cortex following treatment with $\alpha_{2\mu}$ nephropathy-inducing agents and thus are refractory to this disease (Alden 1986; Swenberg et al. 1989; Dietrich and Swenberg 1991a,b; Ridder et al. 1990; MacEwen and Vernot 1978; Gaworski et al. 1980, 1981). In addition, mice, who excrete comparable amounts of the low-molecular-weight mouse urinary protein (MUP) having the greatest amino acid sequence homology to $\alpha_{2\mu}$ (approximately 90%; Borghoff et al. 1990) do not develop the protein-related nephropathy or renal tumors following chronic exposure to d-limonene and other $\alpha_{2\mu}$ nephropathy-inducing agents (National Toxicology Program 1983, 1986a,b, 1987a,b, 1990; Short et al. 1989b). Recent experiments have shown that mice do not develop hyaline droplet nephropathy because MUP does not bind ligands such as d-limonene-1,2-oxide, and because MUP is not resorbed by mouse kidney (Lehman-McKeeman and Caudill 1992a).

Recently, a sex-linked human protein of similar size was identified in urine from patients with renal disease (Bernard et al. 1989, 1990). This protein has been named urine protein 1 and has been called the human equivalent of $\alpha_{2\mu}$ (Bernard et al. 1989). This reference has led to considerable confusion and miscitation. Jackson et al. (1988) purified and partially sequenced human urine protein 1 and determined that it is related to rabbit uteroglobin, not $\alpha_{2\mu}$. It is also present in human urine at concentrations four to five orders of magnitude less than that of $\alpha_{2\mu}$ in male rat urine (EPA 1991).

In an elegant series of experiments, Lehman-McKeeman and Caudill (1992b) have shown that the human members of the $\alpha_{2\mu}$-globulin superfamily and β-lactoglobulin do bind their natural ligands, but not ligands that cause $\alpha_{2\mu}$-globulin nephropathy. In addition, human protein 1 did not bind d-limonene-1,2-oxide or 2,4,4-trimethyl-2-pentanol. Thus, the interaction of $\alpha_{2\mu}$-globulin with agents that cause hyaline droplets, as well as the amount of this protein excreted in the urine, are truly unique to the male rat.

4.4 Epidemiology

No systematic epidemiology studies on nephropathy have been carried out in populations with elevated exposures to industrial chemicals that are known to induce binding to $\alpha_{2\mu}$-globulin. The only human situations studied in which exposure to these chemicals occurred are associated with gasoline. Based on six available studies, an IARC Working Group classified the evidence in humans of carcinogenicity for gasoline as "inadequate" (IARC 1989).

An important limitation of most of the epidemiology studies of renal cancer and gasoline exposure is that hydrocarbon exposures were not quantitatively assessed. In many of the studies, exposure assessment did not go beyond ascertaining that an individual had been employed by a petroleum company or in a refinery. An additional complication of the relevance of these studies to the $\alpha_{2\mu}$ mechanism is that the exposures are to complex mixtures. Two of these studies indicate (McLaughlin 1984;

Siemiatycki et al. 1987) a small increase in relative risk for kidney cancer in subgroups potentially exposed to gasoline. A recently published study (Partanen et al. 1991) found an elevated risk and exposure–response relationship for kidney cancer and gasoline exposure. Summaries of these studies follow.

McLaughlin (1984) found an elevated odds ratio (OR) for occupational exposure to "petroleum, tar, and pitch products" (OR 1.7; 95% confidence interval, CI, 1.0–2.9) in men when adjusted for the confounding factors of age and cigarette smoking. In a subsequent, more detailed analysis of this grouping (McLaughlin et al. 1985) no overall association (OR 1.0; 95% CI 0.7–1.4) was observed for renal cell cancer and employment in a range of occupations with potential for exposure to petroleum products. In a trend analysis, slight risk was associated with duration of employment among service station attendants (OR 1.2; 95% CI 0.6–2.3). The most consistent finding in these studies (McLaughlin and Schuman 1983; McLaughlin 1984) was an associated with cigarette smoking.

Siemiatycki et al. (1987) conducted a population-based case-referent study in Montreal on cancer associations with estimated exposure to 300 chemical materials of which 12 were petroleum-derived liquids. These various mixtures included automotive and aviation gasolines, and distillate jet fuel. No statistically significant risk of renal cancer was found with exposure to automotive gasoline (OR 1.2; 90% CI 0.8–1.6). Statistically significant elevations, however, were noted at the 90% confidence level with exposure to aviation gasoline (OR 2.6; 90% CI 1.2–5.8) and to jet fuel (OR 2.5; 90% CI 1.1–5.4). Aviation gasoline differs in composition from the automotive counterpart by its high content of alkylate naphthas, constituted mainly of branched alkanes (Siemiatycki et al. 1987). Six of the seven cases of exposure to aviation gasoline also involved exposure to jet fuel, making it difficult to distinguish to a unique exposure. In-depth analyses of the two associations using logistic regression methods indicated, however, a greater role for aviation gasoline than for jet fuel.

Partanen et al. (1991) reviewed 672 cases of renal cell adenoma for a case-control study. However, due to a poor participation response only 338 sets of cases and controls were ultimately included for analysis – thereby limiting the interpretability of the analysis. The investigators collected lifelong job histories and translated them into indicators of industry, occupation, and estimated occupational exposure. Elevated risk of kidney cancer was found for a history of employment in white collar occupations; the printing industry; the chemical industry; the manufacturing of metal products, mail, telephone, and telegraph services; and iron and metal work. An elevated risk (OR 1.2; 95% CI 0.6–2.5) was found for gasoline exposure. An exposure–response relationship was observed for increasing exposure to gasoline. A potential confounder, discussed in the paper, is the fact that gasoline used in Finland, especially in the past, contained tetraethyl lead. Lead was associated with increased kidney cancer (OR 2.9; 95% CI 0.5–16.1). In addition, the sparseness of the gasoline exposure estimates required the investigators to develop arbitrary exposure categories which may have maximized the dose–response relationship.

Wong and Raabe (1989) conducted a quantitative meta-analysis by cancer site of petroleum industry employees from the US, Canada, UK, Europe, Australia, and Japan, critically reviewing almost 100 published and unpublished epidemiologic reports. Standardized mortality ratios observed for kidney cancer in the industry as a whole were similar to those for the general population. Results from refinery studies ranged from nonsignificant deficits to nonsignificant excesses. However, the possibility of an elevated kidney cancer risk was raised for one specific group within the industry. Drivers among British distribution workers showed borderline significance for excess kidney cancer mortality. These authors concluded that additional data, particularly involving exposure to downstream gasoline, are needed to resolve the issue. Thus, while small risks cannot be excluded based on specific job categories, the association between human kidney cancer and quantitative exposure to petroleum distillates, if there is one, must be small. Reviews by IARC (1989) and the US EPA (1991) have come to similar conclusions.

In summary, a detailed understanding of the mechanisms involved in $\alpha_{2\mu}$ nephropathy and renal carcinogenesis has been elucidated by investigating several chemicals in various animal, biochemical, and molecular modeling systems. All of the data are consistent with the hypothesis that reversible binding of chemicals or their metabolites to this abundant male rat-specific protein is causally related to the induction of disease. Alternative hypotheses have been proposed, but are not supported by data (Melnick 1992). There are no pathologic nor epidemiologic data available that suggest that humans are susceptible to a similar disease process. The equivocal data on human renal cancer and exposure to petroleum distillates do not address the hypothesis, since a causal association is doubtful and the exposures were to complex mixtures containing hundreds of chemicals. Our present understanding of this disease process strongly suggests that it is unlikely that nongenotoxic chemicals that have been shown to only induce renal tumors in male rats via this mechanism pose a carcinogenic risk to humans.

References

Akerstroem B, Landin B (1985) Rat alpha$_1$-microglobulin: purification from urine and synthesis by hepatocyte monolayers. Eur J Biochem 146:353–358

Akerstroem B, Loedgberg L, Babiker-Mohamed H, Lohmander S, Rask L (1987) Structural relationship between α_1-microglobulin from man, guinea-pig, rat, and rabbit. Eur J Biochem 170:143–148

Alden CL (1986) A review of the unique male rat hydrocarbon nephropathy. Toxicol Pathol 14:109–111

Bernard AM, Lauwerys RR, Noel A, Vandeleene B, Lambert A (1989) Urine protein 1: a sex-dependent marker of tubular or glomerular dysfunction. Clin Chem 35:2141–2142

Bernard AM, Roels H, Cardenas A, Lauwerys R (1990) Assessment of urinary protein 1 and transferrin as early markers of cadium nephrotoxicity. Br J Ind Med 47:559–565

Borghoff SJ, Short BG, Swenberg JA (1990) Biochemical mechanisms and pathobiology of alpha$_{2\mu}$-globulin nephropathy. Annu Rev Pharmacol Toxicol 30:349–367

Borghoff SJ, Miller AB, Bowen JP, Swenberg JA (1991) Characteristics of chemical binding to $\alpha_{2\mu}$-globulin in vitro – evaluating structure-activity relationships. Toxicol Appl Pharmacol 107:228–238

Bruner RH (1984) Pathologic findings in laboratory animals exposed to hydrocarbon fuels of military interest. In: Mehlman MA, Hemstreet GP, Thorpe JJ, Weaver NK (eds) Advances in modern environmental toxicology, vol Vll. Renal effects of petroleum hydrocarbons. Princeton Scientific, Princeton, NJ, pp 133–140

Bruner RH (1986) Chronic consequence of alpha$_{2\mu}$-globulin nephropathy: review. Proceedings of the toxicology forum, Aspen, Colorado, July 15, pp 161–166

Charbonneau M, Strasser J, Lock EA, Turner MJ, Swenberg JA (1989) Involvement of reversible binding to $\alpha_{2\mu}$-globulin in 1,4-dichlorobenzene-induced nephrotoxicity. Toxicol Appl Pharmacol 99:122–132

Chatterjee B, Demyan WF, Song CS, Garg BD, Roy AK (1989) Loss of androgenic induction of $\alpha_{2\mu}$-globulin gene family in the liver of NIH Black rats. Endocrinology 125:1385–1388

Dieppe PA, Doyle DV, Burry HC (1978) Renal damage during treatment with antirheumatic drugs. Br Med J 2:664

Dietrich DR, Swenberg JA (1991a) NCI-Black-Reiter (NBR) male rats fail to develop renal disease following exposure to agents that induce $\alpha_{2\mu}$-globulin ($\alpha_{2\mu}$-G) nephropathy. Fundam Appl Toxicol 16:749–762

Dietrich DR, Swenberg JA (1991b) The presence of alpha$_{2\mu}$-globulin is necessary for d-limonene promotion of male rat kidney tumors. Cancer Res 51:3512–3521

Ekstrom RC, Hoekstra WG (1984) Investigation of putative androgen like activity of alpha$_{2\mu}$-globulin in castrated and estrogen treated male rats. Proc Soc Exp Biol Med 175:491–496

EPA (1991) Alpha$_{2\mu}$-globulin: association with chemically-induced renal toxicity and neoplasia in the male rat. EPAJ625/3-91/091F, Risk Assessment Forum, Environmental Protection Agency,Washington, DC

Gaworski CL, Leahy HF, Brunner HF (1980) Subchronic inhalation toxicity of decalin. In: Proceedings of the 10th conference on environmental toxicology. AFAMRL-TR-121 (ADA086341). Aerospace Medical Research Laboratory, Wright Patterson Airforce Base, OH

Gaworski CL, Haun CC, MacEwan JD (1981) A ninety day inhalation study of decalin. Toxicologist 1:276

Gross V, Steube K, Tran-Thi T, Haeusinger D, Legler G, Decker K, Heinrich PC, Gerok W (1987) The role of N-glycosylation for the plasma clearance of rat secretory proteins. Eur J Biochem 162:83–88:

Gross V, Heinrich PC, vom Berg D, Steube K, Andus T, Tran-Thi T, Decker K, Gerok W (1988) Involvement of various organs in the initial plasma clearance of differently glycosylated rat liver secretory proteins. Eur J Biochem 173:653–659,

Halder CA, Warne TM, Hatoum NS (1984) Renal toxicity of gasoline and related petroleum naphthas in male rats. In: Mehlman MA, Hemstreet GP, Thorpe JJ, Weaver NK (eds) Renal effects of petroleum hydrocarbons. Princeton Scientific, Princeton, NJ, pp 73–87

Halder CA, Holsworth CE, Cockrell BY, Piccirillo VJ (1985) Hydrocarbon nephropathy in male rats: identification of the nephrotoxic components of unleaded gasoline. Toxicol Ind Health 1:67–87

IARC (1987) IARC monographs on the evaluation of carcinogenic risks to humans, suppl 7. Overall evaluation of carcinogenicity: an updating of IARC monographs volumes 1 to 42. IARC, Lyon, France

IARC (1989) IARC monographs on the evaluation of carcinogenic risks to humans, vol 46. Diesel and gasoline engine exhausts and some nitroarenes. IARC, Lyon, France

Jackson PJ, Turner R, Keen JN, Brooksbank RA, Cooper EH (1988) Purification and partial amino acid sequence of human urine protein 1. Evidence for homology with rabbit uteroglobulin. J Chromatogr 452:359–367

Lehman-McKeeman LD, Caudill D (1992a) Biochemical basis for mouse resistance to hyaline droplet nephropathy: lack of relevance of the $\alpha_{2\mu}$-globulin protein superfamily in this male rat-specific syndrome. Toxicol Appl Pharmacol 112:214–221

Lehman-McKeeman LD, Caudill D (1992b) $\alpha_{2\mu}$-Globulin is the only member of the lipocalin protein superfamily that binds to hyaline droplet inducing agents. Toxicol Appl Pharmacol 116:170–176

Lehman-McKeeman LD, Rodriguez PA, Takigiku R, Caudill D, Fey ML (1989) d-Limonene-induced male rat specific nephrotoxicity: evaluation of the association between d-limonene and $\alpha_{2\mu}$-globulin. Toxicol Appl Pharmacol 99:250–259

Lehman-McKeeman LD, Rivera-Torres MI, Caudill D (1990) Lysosomal degradation of $\alpha_{2\mu}$-globulin and $\alpha_{2\mu}$-globulin-xenobiotic conjugates. Toxicol Appl Pharmacol 103:539–548

Lock EA, Charbonneau M, Strasser J, Swenberg JA, Bus JS (1987) 2,2,4-Trimethylpentane (TMP)-induced nephrotoxicity. II. The reversible binding of a TMP metabolite to a renal protein fraction containing alpha$_{2\mu}$-globulin. Toxicol Appl Pharmacol 91:182–192

MacEwen JD, Vernot ER (1978) The effects of subchronic exposure of rodents to inhaled decalin vapors. AMRL 78–55 (ADA062138). Aerospace Medical Research Laboratory, Wright-Patterson Air Force Base, OH (Toxic hazards research unit annual report 1978)

McLaughlin JK (1984) Risk factors form a population-based case-control study of renal cancer. In: Mehlman MA, Hemstreet GP, Thorpe JJ, Weaver NK (eds) Advances in modern environmental toxicology, vol Vll. Renal effects of petroleum hydrocarbons. Princeton Scientific, Princeton, NJ, pp 227–244

McLaughlin JK, Schuman LM (1983) Epidemiology of renal cell carcinoma. Rev Cancer Epidemiol 2:170–210

McLaughlin JK, Blot WJ, Mehl ES, Stewart PA, Venable FS, Fraumeni JF (1985) Petroleum-related employment and renal cell cancer. J Occup Med 27:672–674

Melnick RL (1992) An alternative hypothesis on the role of chemically induced protein droplet ($\alpha_{2\mu}$-globulin) nephropathy in renal carcinogenesis. Regul Toxicol Pharmacol 16: 111–125

National Toxicology Program (1983) Carcinogenesis bioassay of pentachloroethane in F344/N rats and B6C3F1 mice. Research Triangle Park, NC, NTP (National toxicology program technical report series, no 232)

National Toxicology Program (1986a) Carcinogenesis bioassay of tetrachloroethylene (perchloroethylene) in F344/N rats and B6C3F1 mice. Research Triangle Park, NC, NTP (National toxicology program technical report series, no 311)

National Toxicology Program (1986b) Carcinogenesis studies of isophorone in F344/N rats and B6C3F1 mice. Research Triangle Park, NC, NTP (National toxicology program technical report series, no 291)

National Toxicology Program (1987a) Carcinogenesis studies of 1,4-dichlorobenzene in F344/N rats and B6C3F1 mice. Research Triangle Park, NC, NTP (National toxicology program technical report series, no 319)

National Toxicology Program (1987b) Carcinogenesis studies of dimethylmethylphosphonate in F344/N rats and B6C3F1 mice. Research Triangle Park, NC, NTP (National toxicology program technical report series, no 323)

National Toxicology Program (1990) Toxicology and carcinogenesis studies of d-limonene (CAS No. 5989-27-5). Research Triangle Park, NC, NTP (National toxicology program technical report series, no 347)

Olson MJ, Johnson JT, Reidy CA (1990) A comparison of male rat and human urinary proteins: implications for human resistance to hyaline droplet nephropathy. Toxicol Appl Pharmacol 102:524–536

Partanen T, Heikkila P, Henberg S, Kauppinen T, Moneta G (1991) Renal cell cancer and occupational exposure to chemical agents. Scand J Work Environ Health 17:231–239

Pervaiz S, Brew K (1987) Homology and structure-function correlations between alpha$_1$-acid glycoprotein and serum retinol-binding protein and its relatives. FASEB J 1:209–214

Pevsner J, Reed RR, Feinstein PG, Snyder SH (1988) Molecular cloning of odorant-binding protein: member of a ligand carrier family. Science 241:336–339

Read NG, Astbury PJ, Morgan RJI, Parsons DN, Port CJ (1988) Induction and exacerbation of hyaline droplet formation in the proximal tubular cells of the kidneys from male rats receiving a variety of pharmacological agents. Toxicology 52:81–101

Ricca GA, Taylor JM (1981) Nucleotide sequence of rat α_1-acid glycoprotein messenger RNA. J Biol Chem 256:11199–11202

Ridder GM, Von Bargen EC, Alden CL, Parker RD (1990) Increased hyaline droplet formation in male rats exposed to decalin is dependent on the presence of $\alpha_{2\mu}$-globulin. Fundam Appl Toxicol 15:732–743

Roy AK, Neuhaus OW (1966) Proof of hepatic synthesis of a sex-dependent protein in the rat. Biochim Biophys Acta 127:82–87

Roy AK, Neuhaus OW, Harmison CR (1966) Preparation and characterization of a sex-dependent rat urinary protein. Biochim Biophys Acta 127:72–81

Short BG, Burnett VL, Swenberg JA (1986) Histopathology and cell proliferation induced by 2,2,4-trimethylpentane in the male rat kidney. Toxicol Pathol 14:194–203

Short BG, Burnett VL, Cox MG, Bus JS, Swenberg JA (1987) Site-specific renal cytotoxicity and cell proliferation in male rats exposed to petroleum hydrocarbons. Lab Invest 25:564–577

Short BG, Burnett VL, Swenberg JA (1989a) Elevated proliferation of proximal tubule cells and localization of accumulated $\alpha_{2\mu}$-globulin in F344 rats during chronic exposure to unleaded gasoline or 2,2,4-trimethylpentane. Toxicol Appl Pharmacol 101:414–431

Short BG, Steinhagen WH, Swenberg JA (1989b) Promoting effects of unleaded gasoline and 2,2,4-trimethylpentane on the development of atypical cell foci and renal tubular cell tumors in rats exposed to N-ethyl-N-hydroxyethylnitrosamine. Cancer Res 49:6369–6378

Siemiatycki J, Dewar R, Nadon L, Gerin M, Richardson L, Wacholder S (1987) Associations between several sites of cancer and twelve petroleum-derived liquids. Scand J Work Environ Health 13:493–504

Strasser J Jr, Charbonneau M, Borghoff SJ, Turner MJ, Swenberg JA (1988) Renal protein droplet formation in male F344 rats after isophorone treatment. Toxicologist 8:136

Swenberg JA, Short BG, Borghoff SJ, Strasser J, Charbonneau M (1989) The comparative pathobiology of $\alpha_{2\mu}$-globulin nephropathy. Toxicol Appl Pharmacol 97:35–46

Vandoren G, Mertens B, Heyns W, Van Baelen H, Rombauts W, Verhoeven G (1983) Different forms of $\alpha_{2\mu}$-globulin in male and female rat urine. Eur J Biochem 134:175–181

Watabe T, Hiratsuka A, Isobe M, Ozawa N (1980) Metabolism of d-limonene by hepatic microsomes to non-mutagenic epoxides toward Salmonella typhimurium. Biochem Pharmacol 29:1068–1071

Wong O, Raabe GK (1989) Critical review of cancer epidemiology in petroleum industry employees, with a quantitative meta-analysis by cancer site. Am J Ind Med 15:283–310

5 Nongenotoxic Mechanisms in Thyroid Carcinogenesis

G. Thomas

5.1 Introduction

Nongenotoxic mechanisms of thyroid carcinogenesis are of general application and also of potential importance to regulatory toxicology. To understand the way in which administration of xenobiotics can lead to thyroid tumours through a nongenotoxic mechanism it is necessary first to consider the pathophysiology of the thyroid gland.

The thyroid gland normally weighs about 20 mg in a rat of 150 g body weight and is of the order of about 0.01% of the body weight in a number of mammals. It is an unusual endocrine gland as its function is dependent on the dietary supply of a single element, iodine, and its main epithelial component, the follicular cells, are derived embryologically from endoderm. A second, minor epithelial component, the C cells, derive from the neural crest and are responsible for the secretion of calcitonin. Their function is moderated by serum calcium.

Benign thyroid tumours are commonly found in 4%–7% of the human population, but cancer of the thyroid is relatively rare – about 1% of total cancer cases in the USA (Hill et al. 1989). Spontaneous tumours of the follicular cells are rare in rodents; the majority of spontaneous thyroid tumours in aged rats are of C cell origin (Thomas and Williams 1994). However, proliferative thyroid lesions of the follicular cell, including carcinomas, are a relatively common finding in toxicological studies on animals. It is therefore important that the relevance of these animal lesions in terms of human risk assessment is evaluated carefully.

5.2 Structure

The thyroid is a bilobed organ and histologically each lobe is composed of multiple lobules which are composed of numerous spherical follicles. Each follicle is a closed sac lined by epithelial cells. In the unstimulated gland, the follicular epithelial cells are cuboidal in shape and form a single layer, linked at their apical surfaces by tight junctions. The follicular lumen containing the thyroid-specific protein thyroglobulin acts as a reservoir for thyroid hormones. The follicles are supported by a stromal network which is largely composed of fibroblasts and capillaries.

Structure is linked to function, and under the influence of the trophic hormone thyroid stimulating hormone (TSH), the flattened or cuboidal epithelial cells are modified to become columnar. The increase in pinocytosis by the follicular cells stimulated by TSH leads to a depletion of central colloid and the stimulated gland shows small colloid spaces lined by follicular cells containing many colloid droplets.

The ultrastructure of follicular cells also shows changes in response to TSH stimulation. In the normal animal there is an ordered arrangement of cuboidal cells, with apical microvilli, subapical granules and in the body of the cell prominent rough endoplasmic reticulum (ER), mitochondria and occasional lysosomes and phagolysosomes. The iodination of thyroglobulin takes place at the microvillar surface, the subapical granules represent newly synthesised thyroglobulin, the phagolysosomes represent colloid droplets fused with lysosomes and in the process of digestion of thyroglobulin resulting in release of thyroid hormone. Increased TSH stimulation leads to a distension of the ER which may be considerable, more colloid droplets and phagolysosomes and prominent pseudopods on the apical surface, usually near the very prominent junctional complexes where apical surfaces of adjacent follicular cells meet. These pseudopods, which are cellular processes projecting into the colloid, are responsible for engulfing portions of colloid that then form intracellular colloid droplets.

5.3 Synthesis and Secretion of Thyroid Hormone

Synthesis of the thyroid hormones tri-iodothyronine (T3) and tetra-iodothyronine (thyroxine, T4) is dependent upon the dietary supply of iodine. Thyroid hormones are extremely important in the maintenance of metabolic homeostasis and complex mechanisms have evolved to conserve iodide within the thyroid. Synthesis can be divided into four steps:

1. Active uptake into follicular cells. Uptake is effected by the iodide pump. Under normal conditions, the thyroid may concentrate iodide up to about 50-fold its concentration in blood. The pump mechanism can be blocked by anions of a similar size and charge, for example thiocyanate and perchlorate (Wyngaarden et al. 1953). Concurrent uptake of potassium also occurs and drugs which interfere with this process, such as the cardiac glycosides, may also indirectly influence thyroid hormone production (Wolff and Maurey 1961). The active pump mechanism is not unique to the thyroid, similar mechanisms operate in the breast and salivary gland. However, binding of iodide within these tissues does not occur.

2. Oxidation of iodide and formation of iodotyrosines. This reaction is
carried out at the apical border of the follicular cell and involves
two thyroid-specific proteins, thyroglobulin and thyroid peroxidase.
Thyroglobulin is a complex glycoprotein made up of two identical
subunits each with a molecular mass of 330 kDa. It contains specialised tyrosyl residues which are iodinated under the influence of
thyroid peroxidase to give rise to mono-iodotyrosine (MIT) and di-
iodotyrosine (DIT). Thyroid peroxidase can be inhibited by a number of compounds, including the thioureas and ATA (Taurog 1976).
3. Coupling of iodotyrosines to give iodothyronines. This reaction
takes place within the thyroglobulin molecule and is also believed
to be catalysed by thyroid peroxidase. MIT and DIT are coupled to
give rise to T3 and T4 which remain bound to thyroglobulin within
the colloid.
4. Resorption of thyroglobulin. Thyroglobulin uptake from colloid
takes place in the form of resorption of colloid droplets. The rate at
which this resorption takes place is modulated by the level of TSH
stimulation. Monovalent cations such as lithium can inhibit this reaction and give rise to a goitre, which is characterised by colloid
retention (Lazarus 1986). Fusion of the colloid droplet with a lyso-
some leads to degradation of thyroglobulin and release of T4, T3,
MIT and DIT. T3 and T4 are released into the circulation, where
they are bound to plasma proteins. A specialised carrier protein, T4
binding protein, is present in humans but absent in rodents.

In order to conserve iodide, MIT and DIT are deiodinated by thyroid
deiodinase. This enzyme can be inhibited by nitrotyrosines (Stanbury
and Morris 1958). More T4 than T3 is produced by the thyroid. In rats
the ratio is of the order of 4:1, in humans it is between 10 and 20:1,
although the exact ratio varies with iodide intake. T4 has the longer
half-life, but T3 is the more potent on target tissues. The half-life of T4
varies between species; in the rat it is 12–24 h compared to 6–7 days in
the human (Thomas and Bell 1982).

5.4 Metabolism of Thyroid Hormone

Peripheral metabolism of T4 occurs either through outer ring deiodination by the enzyme 5' deiodinase I, which yields the active hormone T3, or by inner ring deiodination which yields reverse T3 (rT3), for which there is no known biological function (Refetoff and Larsen 1989). There are two types of deiodinase: the selenium-containing enzyme deiodinase I, which is widely distributed, but is particularly important in kidney and liver, and deiodinase II, which is restricted in its distribution, found particularly in the pituitary and brain, with T4 as its preferred substrate (Berry et al. 1991). Because of the importance of the trace element selenium for function of deiodinase I, it is thought that selenium deficiency in areas of endemic goitre may further exacerbate the iodide deficiency (Thilly et al. 1993). Propylthiourea can inhibit deiodinase I (Escobar del Rey and Morrelae de Escobar 1961) but not deiodinase II (see Kohrle et al. 1991 for a review on the characteristics of enzymes involved in intracellular pathways of iodothyronine metabolism).

Degradation of thyroid hormones occurs primarily in the liver and involves conjugation with either glucuronic acid (mainly T4) or sulphate (mainly T3). The resulting conjugates are excreted from the bile into the intestine. Many xenobiotics interfere with thyroid hormone metabolism and hepatic elimination. These include phenobarbital (McClain 1989), diproteverine (Flack et al. 1989), polychlorinated biphenyls (Bastomsky 1974), a leukotriene D4 antagonist (L649923; Saunders et al. 1988) and pentachloronitrobenzene (PCNB; Story et al. 1993) to name but a few. It is thought that most of these compounds act by induction of hepatic uridine diphosphate glucuronyl transferase (UDP-Gt) in the rat, leading to appearance of glucuronidated T4 in the bile and increased bile flow rate.

Once in the intestine, a portion of the conjugated thyroid hormone is hydrolysed, and the free hormone reabsorbed into the blood. This reabsorption process can be inhibited by cholestyramine (Northcutt et al. 1969). The remaining conjugates are excreted in the faeces.

Thyroid hormone synthesis and metabolism is a complex process and xenobiotics can alter normal thyroid physiology at several points (see Table 1). Their effects feed back on the growth and function of the thyroid largely through pituitary control.

Table 1. Principle steps of thyroid hormonogenesis and catabolism which are affected by xenobiotics

Step in thyroid hormone synthesis or metabolism	Example of xenobiotic	Reference
Uptake of iodide via iodide pump	Perchlorate, thiocyanate	Wyngaarden et al. (1953)
Concurrent uptake of potassium	Ouabain	Wolff and Maurey (1961)
Organification of iodide	Propylthiourea, methimazole	Taurog (1976); Kohrle et al. (1991)
Coupling of MIT and DIT	Propylthiourea, methimazole	Taurog (1976); Kohrle et al. (1991)
Resorption of colloid	Lithium	Lazarus (1986)
Deiodination of MIT and DIT	Nitrotyrosines	Stanbury and Morris (1958)
Deiodination of T4	Propylthiourea	Escobar del Rey and Morreale de Escobar (1961)
Enzymatic conjugation	PCBs, phenobarbital	Bastomsky (1974); McClain (1989)
Resorption from the intestine	Cholestyramine	Northcutt et al. (1969)

The compounds listed are examples only. The reader is recommended to refer to Hill et al. (1989) and Atterwill et al. (1992) for a more comprehensive list.

MIT, mono-iodotyrosine; DIT, di-iodotyrosine; PCB, polychlorinated biphenyl.

5.5 Thyroid–Pituitary Interaction

The main regulator of both function and growth of the thyroid follicular epithelium is TSH, which is produced by the thyrotrophs of the anterior pituitary. TSH acts on key stages in thyroid hormone synthesis (the efficiency of the iodide pump, thyroglobulin and thyroid peroxidase synthesis, colloid resorption) via the cyclic adenosine monophosphate (cAMP) pathway (van Sande and Dumont 1973). A higher tier of control is exerted by the hypothalamus which secretes thyroptropin releasing hormone (TRH), which stimulates TSH release from the thyrotrophs in the pituitary. T3 binds to nuclear receptors in the thyrotroph and the level of T3 receptor occupancy is inversely related to the secretion of TSH. The circulating level of T4 plays an important role in this feedback mechanism as the thyrotrophs contain 5' deiodinase II, which converts T4 to T3. There is therefore an inverse relationship between the levels of circulating T4 and TSH. This system is extremely sensitive to fluctuations in T4, induced either by alteration in intrathyroidal synthetic pathways or an increase in loss of T4 from the plasma by alteration of liver metabolic pathways. It serves to increase TSH secretion by the pituitary before any detrimental effects due to a serious decrease in the level of the more active hormone T3.

5.6 Effect of Iodide on Thyroid Hormone Synthesis

When the iodide supply is insufficient, relatively more MIT than DIT is formed on thyroglobulin. This results in a decreased ratio of T4 to T3 production. Because T3 is the more potent hormone on target tissues, this mitigates the deficiency of thyroid hormone as far as target sites are concerned. In addition, the fact that the pituitary thyrotroph is more sensitive to the level of circulating T4 than T3, the reduction in T4 production activates the pituitary feedback mechanism, leading to a compensatory rise in TSH before there is any significant detrimental effect on thyroid hormone supply to target organs. TSH stimulation increases the efficiency of the iodide pump and formation of T3 and T4 as well as increasing uptake and release of stored hormone. Iodide also has direct effects – in excess it decreases the efficiency of the iodide pump, an effect which is believed to be mediated by an oxidised iodide moiety (Nagataki and Ingbar 1986).

Often iodide insufficiency or xenobiotic effects are transient and do not produce a marked decrease in thyroid hormone production. These effects can easily be countered by a slight and short rise in TSH leading to an increase in function only of the gland. However, when there is a significant and prolonged decrease in thyroid hormone levels, a persistent rise in TSH is observed. This then leads to thyroid growth.

5.7 TSH and Normal Thyroid Growth

In addition to it effects on function, TSH is the major growth factor for the thyroid follicular epithelium. TSH stimulation initially produces an increase in thyroid weight. The follicles become lined by tall columnar cells and there is a depletion of colloid from the centre of the follicles.

Sustained elevated TSH levels produced by administration of the thyroid peroxidase inhibitor aminotriazole (ATA) does not lead to continued growth of the rat thyroid. Careful morphometric analysis has shown that the growth curve is triphasic; initially growth of both the follicular cells and the surrounding stroma is rapid, leading to a coordinated increase in the number of stromal and epithelial cells. After about 3 months a drop in mitotic activity is observed, leading to a plateau phase which lasts about 5 months. Finally, thyroid weight increases again due to the production of proliferative lesions within the gland (Wynford-Thomas et al. 1982a). The decrease in the mitotic response to TSH is not mirrored by a decrease in the follicular cells functional response to TSH, measured by the efficiency of the iodide pump, and is not due to a fall in the level of TSH produced by the pituitary. These results suggested that there was a mechanism by which the follicular cell uncoupled its growth and functional responses to TSH (Wynford-Thomas et al. 1982b). Removal of the goitrogen to produce a decrease in TSH levels followed by a reintroduction of goitrogen feeding 28 days later did not result in a second burst of mitotic activity in the follicular cells, despite an increase in TSH levels (Wynford-Thomas et al. 1982c). However, the lack of responsiveness of the follicular epithelium with respect to growth in response to TSH was shown to be at least partly stimulus specific. Incision of the thyroid isthmus in rats maintained on goitrogen for 3 months produced localised reparative growth near to the site of injury (Wynford-Thomas et al. 1985).

Studies of growth control in vitro using thyroid cell lines and primary cultures have been used to explain the in vivo observations. Primary cultures of thyroid follicular cells do not survive more than a few days or weeks in vitro, perhaps due to the growth-limiting mechanism. At one time, it was even claimed that TSH did not produce growth in thyroid cells in culture (Westermark et al. 1979). Using a TSH-responsive immortal rat thyroid cell line, it was subsequently shown that a cocktail of five other growth factors was required to elicit growth in response to TSH (Ambesi-Impiombato et al. 1980). It is now recognised that a single growth factor, insulin-like growth factor (IGF)1, permits the growth response to TSH. Initially, supraphysiological levels of insulin were shown to stimulate growth in response to TSH, but it is now recognised that this response results from stimulation of the IGF1 receptor. Many species share this growth factor requirement, human (Williams et al. 1987,1988), rat (Smith et al. 1986) and dog (Roger et al. 1983). The pig is, however, an exception (Gartner and Greil 1986). Our own work, using in situ hybridisation for detection of mRNA and immunocytochemistry for detection of peptide, has shown that IGF1 is produced by murine thyroid follicular cells in vivo. In the normal thyroid, there is intra and interfollicular heterogeneity of expression of both mRNA and peptide, but short-term goitrogen treatment leads to a marked rise in IGF1 mRNA content of the follicular cells and a loss of inter- and intrafollicular heterogeneity (Thomas et al. 1994). The fact that exogenous administration of IGF1 is required for growth in vitro suggests that, in vivo, IGF1 production by follicular cells may be elicited by an extrafollicular cell stimulus, perhaps originating from the stroma. Stromal and epithelial cells grow in a coordinated fashion in response to TSH stimulation in vivo, which suggests there is communication between the two cell types. However, the role of IGF1 is complicated by the fact that its activity may be modulated by the presence of its associated binding proteins.

Growth factors other than IGF1 have also been shown to produce follicular cell growth in vitro: fibroblast growth factor (FGF; Roger and Dumont 1984) and IGF2 (Maciel et al. 1988), and epidermal growth factor (EGF) which stimulates proliferation, but not differentiation (Roger and Dumont 1984). Other factors such as transforming growth

factor beta (TGF-β; Grubeck-Loewenstein et al. 1989) may be inhibitory for follicular cell growth.

Iodide status can also modify the response to TSH stimulation in
vivo. The growth response to exogenously administered TSH in hypophysectomised rats is greater in iodide-deficient than in iodide replete
animals (Bray 1968). This effect may be mediated via an arachidonic
acid metabolite (Pisarev et al. 1986) or other organified compound
(Becks et al. 1988).

Physical factors may stimulate growth – the burst of mitotic activity
observed by Many et al. (1983) on refeeding iodide to iodide-deficient
mice could have resulted from growth in response to distension of
follicles.

Finally, growth can occur in the absence of TSH as evidenced by the
growth of thyroid autografts in hypophysectomised animals (Williams
and Doniach 1963). Reparative growth seen in the TSH-suppressed rat
in response to injury may also operate through a TSH-independent
mechanism (Wynford-Thomas et al. 1985).

5.8 Thyroid Carcinogenesis

In the human, thyroid tumours can be divided into two types of differentiated tumour, follicular and papillary, and one type of undifferentiated
tumour, anaplastic carcinoma. Anaplastic carcinomas frequently do not
express the thyroid-specific genes thyroid peroxidase and thyroglobulin
and this may be due to loss of expression of tissue-specific transcription
factors TTF1 and Pax 8 (Heldin and Westermark 1991). It is likely that
anaplastic lesions are a result of clonal progression from pre-existing
differentiated lesions. The distinction between the two differentiated
types of thyroid tumour, follicular and papillary, is based not only on
morphology, but also upon differing cytology, method of invasion and
spread, pattern of oncogene involvement and relationship to iodide
deficiency, papillary lesions being more common in iodide-rich areas
(Williams et al. 1977).

Distinction between types of lesions in the rodent is not so easy, and
frequently tumours are a mixture of papillary and follicular architecture.
Although, interestingly, the papillary areas do not show the characteristic grooved nuclei of the human papillary tumour. However, a

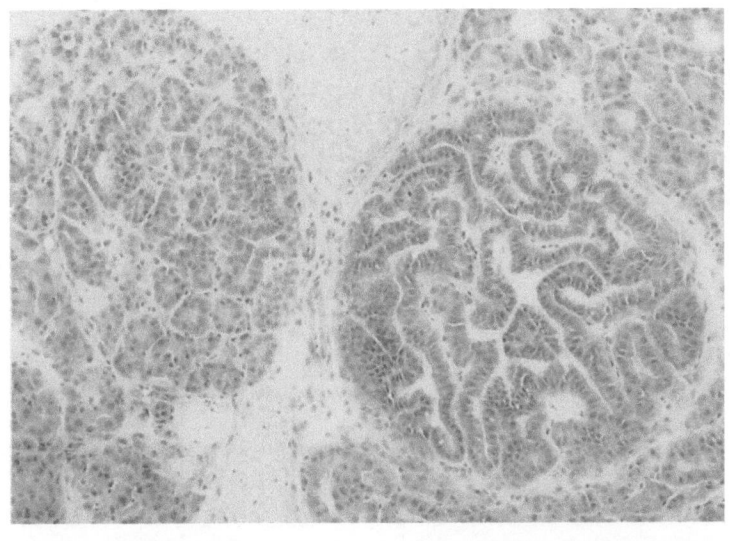

a

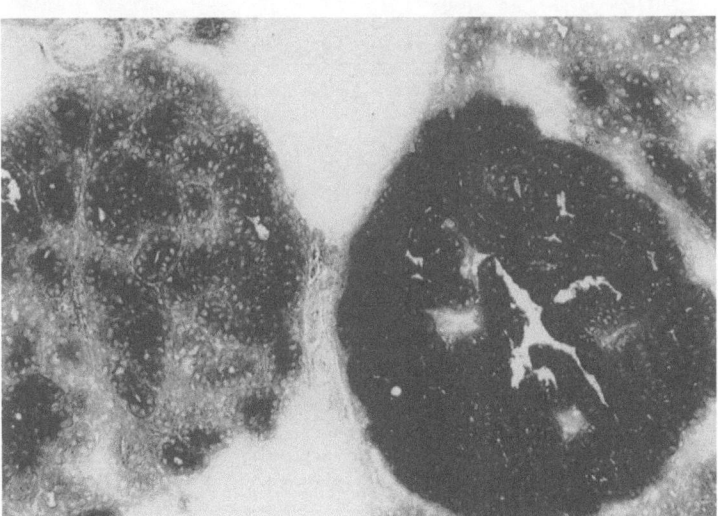

b

Fig. 1a–d. Legend on p. 91

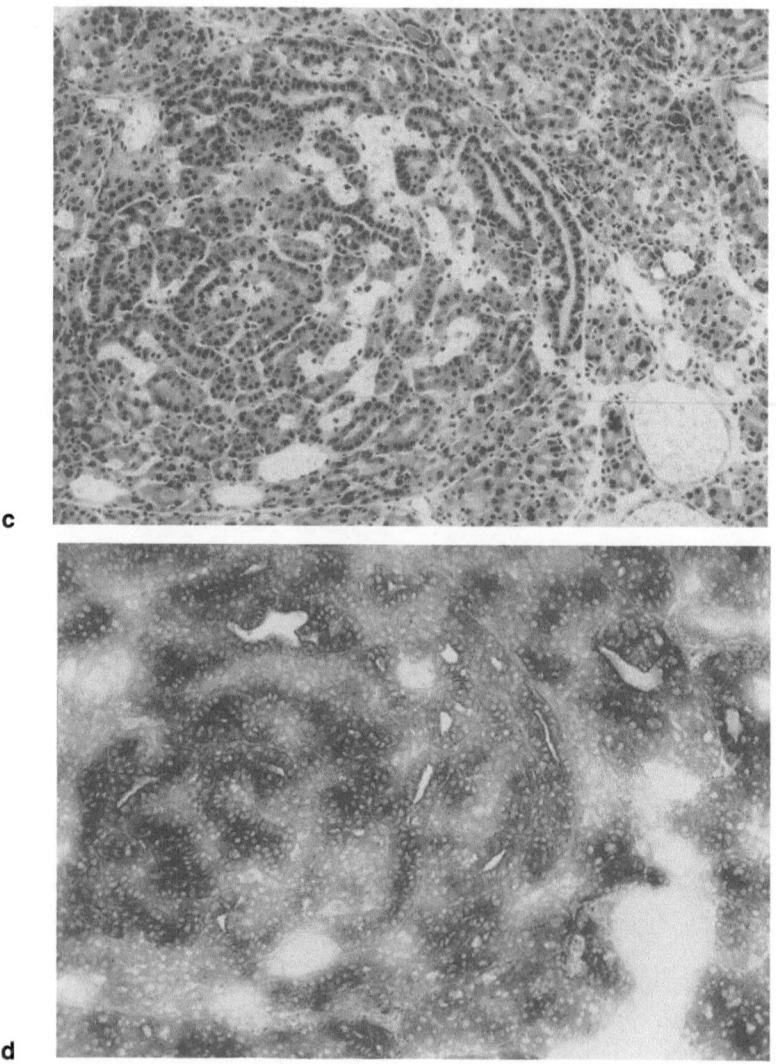

c

d

Fig. 1a–d. Legend on p. 91

morphological distinction between two morphological types of benign lesion can be made in the mouse. The adenoma is characteristically composed of abnormal follicular or papillary architecture, with crowded basophilic epithelial cells and a scanty stromal component (Fig. 1a). In contrast, nodules retain some normal follicular architecture, with similar cytology to the non-neoplastic thyroid, but show an excessive stromal component (Fig. 1c).

5.9 Congenital Thyroid Dyshormonogenesis

Congenital thyroid dyshormonogenesis is rare, but several defects have been identified. The commonest defects are the lack of the iodide pump (Stanbury and Chapman 1960), defective thyroid peroxidase (Stanbury and Hedge 1950) or thyroglobulin (Alexander and Burrow 1970) and the absence of intrathyroid deioidinase (Salvatore et al. 1980). Similar

---

Fig. 1a–d. (p. 89/90) Clonality of the two morphological types of benign tumour induced in mice after prolonged goitrogen administration. **a,b** Adenoma. **a** Frozen H&E stained section from a C3HxGPDX mouse given a single intraperitoneal injection of 131-iodine (3 µCi) at 3 weeks of age and subsequently aminotriazole (0.2%) in the drinking water for 60 weeks. There is a well-demarcated adenoma characterised by abnormal follicular/papillary architecture, with crowded basophilic epithelial cells and a scanty stromal component. **b** Serial section to **a** stained with the enzyme histochemical technique for glucose-6-phosphate dehydrogenase (G6PD). In this heterozygous female, two cellular patterns of enzyme reaction can be seen in the normal background hyperplastic thyroid. Cells in which the mutant deficient G6PD gene is on the active X-chromosome show low enzyme activity, whereas those in which the normal gene is on the active X-chromosome show high levels of G6PD activity. The tumour is entirely composed of cells which express high G6PD activity, showing that it is monoclonal in origin. Magnification × 160. **c,d** Nodule. **c** Frozen H&E stained section from a C3HxGPDX mouse maintained on 1% perchlorate in the drinking water for 60 weeks. The nodule is composed of follicles of mainly normal cellularity and architecture, but there is an increased stromal content within the lesion. **d** Serial section stained with the enzyme histochemical technique for G6PD. Both enzyme phenotypes are clearly visible in both the nodule and in the background hyperplastic thyroid, indicating that the epithelial component of the nodule is of polyclonal origin. Magnification × 200. (From Thomas et al. 1989 by kind permission of the American Society for Investigative Pathology Inc.)

defects have been identified in other large mammals such as Afrikander cattle (van Jaarsveld et al. 1972), merino sheep (Falconer et al. 1965) and a breed of Dutch goat (de Vijlder et al. 1978).

These defects produce a similar pathology in both humans and animals. Initially the thyroid is hyperplastic, which if left untreated becomes increasingly nodular and eventually may result in carcinoma (Vickery 1981). The pathogenesis of this sequence of events is not known, but it appears that the only influence on the production of lesions is elevation of TSH due to operation of the pituitary feedback control mechanism.

5.10 Production of Thyroid Lesions by Dietary Modification

In many ways manipulation of the intake or organification of iodide by diet or by administration of antithyroid drugs mimics the effect of congenital thyroid dyshormonogenesis. Feeding an iodide-poor diet is analogous to a deficiency of iodide supply resulting from an iodide pump defect, and administration of thyroid peroxidase inhibitors produces the same effect as possession of defective thyroid peroxidase protein. Goitre is a frequent finding in both humans and animals in areas of the world which are iodide deficient, and feeding of an iodide-deficient diet alone has been shown to induce a low frequency of thyroid tumours in rats (Axelrad and Leblond 1955; Schaller and Stevenson 1966). However, concomitant administration of a goitrogenic agent can further exacerbate the iodide deficiency by rendering the animal more hypothyroid. Goitrogens occur commonly in the environment both as pollutants and as natural substances in the food we eat (Gaitan 1988). Glucobrassin, a thiocyanate present in cabbage, may play a role in the pathogenesis of goitre in central Europe (Ermans 1978) and cassava, a basic foodstuff in the tropics, is goitrogenic in rats and contains thiocyanate, which inhibits iodide uptake (Ekpechi et al. 1966). Early studies on the induction of goitre and thyroid carcinogenesis used prolonged dietary administration of seeds from the *Brassica* species (Griesbach et al. 1945). It was subsequently found that the active agent was a urea derivative (Gmelin and Virtanen 1960); several other chemically similar compounds have been shown to be both goitrogens and carcinogens in experimental animals.

5.11 Production of Thyroid Tumours by Xenobiotics

Out of a total of 300 chemicals listed in the NCI/NTP database only 21 compounds have been associated with the development of follicular cell tumours. These fall into four groups: thionamides, e.g N,N'-die-thylthiourea; aromatic amines, e.g. 3-amino-4-ethoxyacetanilide and 4,4'-oxydianiline; complex halogenated hydroacarbons, e.g. chlorinated paraffins (C_{12}, 60% chlorine); and one organophosphorous compound, azinphosmethyl. The majority of chemicals fall into the first two groups (13/21), the bulk of the remainder being halogenated hydrocarbons (7/21). Thionamides and aromatic amines are known to exhibit peroxidase inhibition, and some halogenated hydrocarbons have been shown to increase clearance of thyroid hormones from the blood. However, not all members of the same chemical class of compound necessarily produce positive thyroid effects. However, in the majority of cases, agents producing thyroid tumours could be shown to have some effect on thyroid hormone homeostasis (Hill et al. 1989).

Prolonged administration of thyroid peroxidase inhibitors such as the thioureas methimazole or ATA and iodide pump inhibitors such as perchlorate have been shown to produce thyroid tumours, including carcinomas in both mice and rats (for review see Paynter et al. 1988, Hill et al. 1989, Thomas and Williams 1991). In addition, agents which affect the conjugating enzymes of the liver, heavily iodinated compounds such as amiodarone and compounds acting through alteration of hypothalamic–pituitary control, such as theophylline, lead indirectly to an increase in TSH secretion by the pituitary (Atterwill et al. 1992).

There is evidence that hypophysectomy or concurrent T4 administration inhibits both the goitrogenic and carcinogenic effects of such compounds in experimental animals (Jemec 1980), suggesting that the role in pituitary secretion of TSH is of key importance in the generation of thyroid tumours by nongenotoxic agents. However, the genotoxic potential of these agents cannot be completely excluded as thyroid carcinogenesis caused by a known mutagen, radiation, can also be inhibited by suppression of TSH production (Nadler et al. 1970).

5.12 Genotoxic Chemicals and Thyroid Tumorigenesis

5.12.1 Radiation

Radiation plays a dual role in thyroid tumorigenesis, both genotoxic and nongenotoxic. Administration of both 131-iodine (Doniach 1953) and external radiation (Doniach 1957) have been shown to induce tumours in animals. There is also evidence from epidemiological studies of human populations exposed to nuclear fallout from atomic weapons testing that radiation is a thyroid carcinogen in humans (de Groot 1988). It is also becoming increasingly likely that a significant increase in childhood thyroid cancer in Belarus may be the result of nuclear fallout after the Chernobyl accident (Baverstock et al. 1992).

Experimental evidence from rats suggests that radiation does not produce a linear dose–response curve. The dose which gives the highest frequency of thyroid tumours in rats (30 μCi; Doniach 1953) has been shown to lie within a dose range which demonstrably interferes with thyroid function (Maloof et al. 1952). In an elegant experiment using X-irradiation, with shielding of one of the two lobes, Nichols et al. (1965) showed that lesions developed in both lobes, but were more frequent in the unshielded lobe. These results suggest that there was clearly a secondary component to radiation-induced carcinogenesis. Observation that both hypophysectomy (Nadler et al. 1970) and T4 administration (Doniach 1974), and the fact that higher doses of radioiodine which rendered follicular cells incapable of division, inhibited radiation-induced carcinogenesis suggested that the mitotic stimulus provided by TSH stimulation was an important ingredient in the pathogenesis of these lesions. In addition, the induction of thyroid tumours can be further potentiated by subsequent administration of an iodide-poor diet (Axelrad and Leblond 1955) or by goitrogens (Doniach 1953; Thomas and Williams 1991).

This evidence suggests that radiation induces thyroid tumours by both a direct genotoxic mechanism and by nongenotoxic mechanisms involving growth stimulation of the follicular epithelium by TSH.

5.12.2 Chemical Carcinogens

Few compounds which induce thyroid tumours do so by a mechanism which is completely exclusive of TSH involvement. Aromatic amines such as aminofluorene (AAF, Doniach 1950), azo dyes, such as 4,4'-methylene-bis-(N,N-dimethyl)-benzamine (MDBA, Murthy 1980), nitrosamines, e.g. diisopropanolnitrosamine (DIPN, Mohr et al. 1977), N-bis-(2-hydroxypropyl)-nitrosamine (DHPN, Hiasa et al. 1982) and nitrosoureas, e.g. methylnitrosourea (MNU, Tsuda et al. 1983), all exhibit a low frequency of thyroid tumours when administered in multiple doses to rats. The frequency of lesions and their latency can be markedly influenced by subsequent administration of goitrogens, again emphasising the role for a nongenotoxic component in thyroid tumorigenesis.

In addition to combined genotoxic and nongenotoxic agents, the frequency of induced lesions can be increased by combining an antithyroid drug with nongenotoxic agents which demethylate DNA. Our own recent studies showed that 5-azacytidine can potentiate thyroid tumour production by ATA in mice (Thomas and Williams 1992).

5.13 Pathobiology of Thyroid Tumours

Spontaneous tumours of the thyroid follicular cell are rare in experimental animals. This may be due in part to the fact that induction of cancer requires the accumulation of multiple mutations in a single cell. Cell division is required to "fix" a mutation within a cell and under normal circumstances the thyroid has a low mitotic rate. Recent studies using transgenic mice have suggested that when thyroid-specific hyperplasia is induced by coupling the thyroglobulin promoter to either the simian virus 40 T antigen (Ledent et al. 1991) or to the A2 adenosine receptor, which leads to constitutive activation of the cAMP cascade (Ledent et al. 1992), thyroid tumours develop very soon after birth. In fact, it has been almost impossible to establish transmission of the transgene with the SV40T transgene as the mice are born with compression of the trachea and oesophagus by hyperplastic thyroid tissue and develop poorly differentiated thyroid adenocarcinomas before breeding age (Ledent et al. 1991). The A2 receptor transgenics presented with less aggressive thyroid disease, but nevertheless developed hyperplasia and

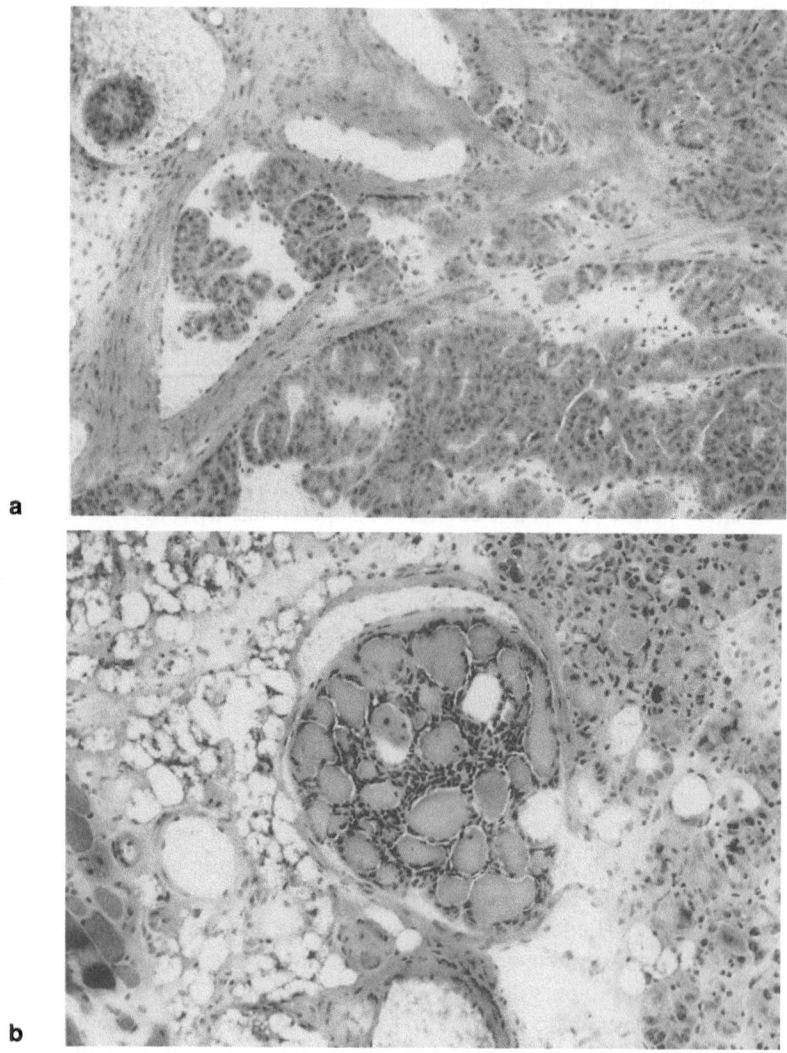

Fig. 2a,b.

neoplasia which caused premature death of the animals (Ledent et al. 1992).

The generation of thyroid tumours by administration of xenobiotics which disturb thyroid hormonogenesis is a more protracted affair than that observed in transgenic mice. Diffuse hyperplasia progresses to formation of benign tumours and eventually in a few animals to carcinomas. A definition of what constitutes the difference between hyperplasia and neoplasia is frequently difficult in pathology. However, one of the key features of neoplasia is clonal progression through acquisition of a series of mutations in a single cell (the word "mutation" is used here to encompass not only base pair changes but also overexpression, inappropriate expression and translocation). Therefore, one possible definition of a neoplastic lesion is that which results from overgrowth of a single clone of cells. However, there is not one single adequate definition of what constitutes a tumour and reversibility of a lesion is frequently used to define that which is and that which is not a tumour.

Prolonged administration of goitrogens to mice results in two types of benign tumour, the adenoma (Fig. 1a) and the nodule (Fig. 1c). Using a mouse model which enables us to study clonality and the cellular level (Thomas et al. 1988), we have shown that adenomas are monoclonal in origin and therefore result from a heritable alteration in growth control of a single cell (Fig. 1b) and nodules are polyclonal (Fig. 1d; Thomas et al. 1989).

Fig. 2a,b. Morphological regression of carcinomas after withdrawal of the goitrogenic agent. **a** Follicular carcinoma induced in a C3HxGPDX mouse by a single intraperitoneal injection of 131-iodine (23 μCi) at 3 weeks of age followed by 46 weeks of oral goitrogen (0.2% aminotriazole/0.5% perchlorate) administration. There is clear evidence of extracapsular vascular invasion. H&E × 160. **b** Section from a C3HxGPDX mouse given a single intraperitoneal injection of 131-iodine (23 μCi) at 3 weeks of age prior to 46 weeks of oral goitrogen (0.2% aminotriazole/0.5% perchlorate) administration. Goitrogen was then withdrawn for 4 weeks prior to sacrifice. There was clear evidence of morphological regression in both tumour within the thyroid (not shown) but also in endothelialised tumour which had invaded extracapsular veins. These results suggest that even invasive tumours may regress on withdrawal of the goitrogenic stimulus and are therefore still dependent on thyroid stimulating hormone for growth. H&E × 300. (From Thomas et al. 1991 by kind permission of Macmillan Press Ltd.)

Addition of a single intraperitoneal injection of a high dose of 131-iodine to a goitrogenic regime in these mice also induced monoclonal adenomas and a small proportion of monoclonal carcinomas (Thomas et al. 1991). However, removal of goitrogen administration resulted in morphological regression of these lesions, including carcinomas with clear extracapsular vascular invasion (Fig. 2 and Thomas et al. 1991). Previous authors had also noticed regression of both adenomas (Todd 1986) and carcinomas and even tissue metastatic to the lung (Dunn 1975, Jemec 1977) after withdrawal of goitrogen. Transplantation of neoplastic tissue from animals maintained on goitrogen to euthyroid animals also resulted in regression, whereas transplantation to hypothyroid animals maintained the malignant phenotype (Matovinovic et al. 1970). Regression results in a loss of cells from the gland, but if goitrogen administration has been prolonged prior to induced regression, the thyroid remains two to three times its normal size and weight, despite a return to normal histological appearance (Greer et al. 1967; Wollman and Breitman 1970). It is not known whether the tumour phenotype would reappear in these animals if goitrogen administration was recommenced, although transplantation experiments suggest that this may be the case (Matovinovic et al. 1970).

These experiments leave us with an interesting conundrum: tumours induced by prolonged goitrogen administration, which presumably result from a nongenotoxic mechanism, are monoclonal in origin, suggesting heritable mutation in growth control of a single cell, but regress when TSH stimulation is withdrawn. We know that the follicular cells' growth response to TSH is limited (Wynford-Thomas et al. 1982a). The first key step in TSH-induced carcinogenesis would logically be loss of this restriction mechanism – presumably a tumour suppressor gene. A second key step must be the development of independence from TSH stimulation. However, when tumours are induced by a regime which involves maximal stimulation of the follicular cell by TSH, it seems unlikely that escape from TSH-mediated control would confer a growth advantage. The third key step must involve evolution of the ability to invade. The cellular mechanisms by which this progression is achieved are the subject of considerable interest.

5.14 Oncogenes in Thyroid Carcinogenesis

Several oncogenes have been implicated in the progression of thyroid neoplasia. However, it is not yet known which particular oncogenes are involved in which particular step, but we can say which are likely to be early and which late events. For example, mutation in the p53 gene is likely to be involved in the progression from adenoma to carcinoma in the human, as mutations have not yet been found in early lesions (Donghi et al. 1993; Fagin et al. 1993; Nakamura et al. 1993). Mutation of the *ras* oncogenes is, however, likely to occur at an early stage in the formation of follicular adenomas in both man and animals, and there is a suggestion that involvement of individual members of the *ras* family may be agent specific (Lemoine et al. 1988, 1989). It is not clear, however, whether *ras* gene mutation allows escape from the growth-desensitising mechanism to TSH or whether their mutation represents a later step.

Other oncogenes which have been shown to be involved in human thyroid neoplasia include c-*met*, which encodes the receptor for hepato-cyte growth factor. C-*met* expression is normally low in follicular cells but is overexpressed in carcinomas of the follicular cell, but not in C cell tumours or follicular adenomas. Initial data indicate that c-*met* involve-ment may therefore represent a late step in tumorigenesis (Di Renzo et al. 1992).

Interestingly, papillary thyroid tumours in humans show a different pattern of oncogene involvement from follicular tumours, which sug-gests that there may be two different pathways involved in the gener-ation of these two different pathological entities. In human papillary tumours translocations of genes encoding tyrosine kinase linked recep-tors appear to play a key role. Approximately 20% of papillary tumours exhibit a translocated *ret* oncogene and a further 20% a translocated *trk* oncogene (Bongarzone et al. 1989). *Ret* is normally expressed in cells of neuroendocrine origin, and mutation of this oncogene has recently been shown to be involved in a rare tumour syndrome, multiple endocrine neoplasia type 2A (MEN 2A) which involves development of medullary carcinoma from the C cells of the thyroid (Mulligan et al. 1993). Some normal human C cells, but not follicular cells, appear to express *ret* (G. Thomas et al., unpublished observations), which suggests that part of the pathogenesis of papillary carcinoma involves activation of a normally silent gene. The ligand for this tyrosine kinase receptor is not yet known, and we do not yet know whether tumours induced in animals by radiation and goitrogen treatment also express *ret*. Tumours induced in animals, while exhibiting similar papillary architecture to that seen in the human, do not share the characteristic nuclear change of human papillary tumours and do not exhibit the same characteristic lymphatic metastatic spread.

5.15 Growth Factor Control of Thyroid Tumours

Many of the oncogenes discussed above encode growth factor receptors; for some, such as c-*met*, the ligand is known, whereas others, such as *ret*, have yet to have a ligand assigned to them. The local production of growth factors is therefore likely to be of great importance in thyroid neoplasia. IGF1 has been shown to be a key factor in normal growth control of the follicular epithelium. In vitro studies have shown that follicular cells cultured from human thyroid adenomas did not require exogenous administration of IGF1 for their growth response to TSH (Williams et al. 1988). Immunocytochemical studies have shown that cells from adenomas appeared to be producing IGF1 peptide themselves, which leads to the suggestion that autocrine production of IGF1 might represent one of the steps in thyroid tumorigenesis (Williams et al. 1989). However, animal studies in vivo have shown that IGF1 is normally produced by the follicular cells of the mitotically active young thyroid, and that administration of goitrogen to adult animals increases both IGF1 mRNA and peptide in the follicular cells (Thomas et al. 1994). Early indications from further studies suggest that IGF1 production by the follicular cells is not maintained at a uniformly high level in animals maintained on goitrogen, but that tumours induced by the standard radiation and goitrogen regime do not exhibit elevated levels of IGF1 production (G. Thomas et al., unpublished observations). However, regulation of IGF1 action is complicated by the existence of several binding proteins which modify its action and further studies of the distribution of binding proteins in differing physiological states will be required before the role of IGF1 in thyroid tumorigenesis is fully understood.

We are therefore left with a series of pieces of a molecular biological puzzle which require slotting into our understanding of the pathogenesis of thyroid neoplasia. TSH clearly plays a central role in the generation of thyroid tumours in experimental animals, and in some humans with dyshormonogenesis or who are exposed to chronic iodide deficiency. However, the mechanism by which these tumours develop is not fully understood. There is clearly a place for the involvement of *ras* and other oncogenes in follicular adenoma and for p53 in the development of carcinoma. Oncogene alterations must in some way relate to alteration in growth signalling pathways as many oncogenes, such as *ret*, *trk* and *met*, encode growth factor receptors. It also seems that there are several pathways to tumour production and the method of induction – follicular tumours being more common in iodide-poor regions and papillary tumours being more frequent after radiation exposure and in iodide-rich areas – may lead to a pathway-specific oncogene involvement.

We should also not forget that while, in general principles, production of neoplasia may be similar in humans and animals, species-specific differences do occur. In regulatory terms, it would appear that humans are less sensitive to thyroid tumour induction than animals and it may be possible to apply safety threshold levels (Paynter et al. 1988). However, mechanisms of nongenotoxic carcinogenesis in the rodent thyroid remind us just how complex hormonal homeostasis can be and illustrate that xenobiotics can influence growth and function of endocrine tissues at many different levels.

Acknowledgements. I would like to thank the Medical Research Council for financial support and Professor Sir Dillwyn Williams for many years of patient and helpful discussion.

References

Alexander NM, Burrow GN (1970) Thyroxine biosynthesis in human goitrous cretinism. J Clin Endocrinol Metab 30:308–315

Ambesi-Impiombato FS, Parks LAM, Coon HG (1980) Culture of hormone-dependent functional epithelial cells from rat thyroids. Proc Natl Acad Sci USA 77: 3455–3459

Atterwill CK, Jones C, Brown CG (1992) Thyroid gland II – mechanisms of species-dependent thyroid toxicity, hyperplasia and neoplasia induced by xenobiotics. In: Atterwill CK, Flack JD (eds) Endocrine toxicology. Cambridge University Press, Cambridge, pp 137–182

Axelrad AA, Leblond CP (1955) Induction of thyroid tumors in rats by a low iodine diet. Proc Am Assoc Cancer Res 1:2

Bastomsky C H (1974) Effects of a polychlorinated biphenyl mixture (Arachlor 1254) and DDT on the biliary thyroxine excretion in rats. Endocrinology 95:1150–1155

Baverstock K, Egloff B, Pinchera A, Ruchti C, Williams ED (1992) Thyroid cancer after Chernobyl. Nature 349:21–22

Becks GP, Eggo MC, Burrow GN (1988) Organic iodine inhibits deoxyribonucleic acid synthesis and growth in FRTL-5 thyroid cells. Endocrinology 123:545–551

Berry MJ, Banu L, Chen Y, Mandel SJ, Kieffer JD, Harney JW, Larsen PR (1991) Recognition of UGA as a selenocysteine codon in type I deiodinase requires sequences in the 3' untranslated region. Nature 353:273–276

Bongarzone I, Pierotti MA, Monzini N, Mondellini P, Manenti G, Donghi R, Pilotti S, Grieco M, Santoro M, Fusco A, Vecchio G, Della Porta G (1989) High frequency of activation of tyrosine kinase oncogenes in human papillary thyroid carcinoma. Oncogene 4:1457–1462

Bray GA (1968) Increased sensitivity of the thyroid in iodine-depleted rats to the goitrogenic effects of thyrotrophin. J. Clin Invest 47:1640–1647

de Groot LJ (1988) Radiation and thyroid disease. Ballieres Clin Endocrinol Metab 2:777–791

de Vijlder JJM, van Voorthuizen WF, van Dijk JE, Rijnberk A, Telegaers WHH (1978) Hereditary congenital goiter with thyroglobulin deficiency in a breed of goats. Endocrinology 102: 1214–1222

Di Renzo MF, Olivero M, Ferro S, Prat M, Bongarzone I, Pilotti S, Belfiore A, Costantino A, Vigneri R, Pierotti MA (1992) Overexpression of the c-met/HGF receptor gene in human thyroid carcinomas. Oncogene 7:2549-2553

Donghi R, Longoni A, Pilotti S, Michieli P, Della Porta G, Pierotti MA (1993) Gene p53 mutations are restricted to poorly differentiated and undifferentiated carcinomas of the thyroid gland. J Clin Invest 91:1753-1760

Doniach I (1950) The effect of radioactive iodine alone and in combination with methylthiourea and acetylaminofluorene upon tumour production in the rats thyroid gland. Br J Cancer 4:223–234

Doniach I (1953) The effect of radioactive iodine alone and in combination with methylthiouracil upon tumour production in the rats thyroid gland. Br J Cancer 7:181–202

Doniach I (1957) Comparison of the carcinogenic effect of X-irradiation with radioactive iodine on the rats thyroid. Br J Cancer 11:67–76

Doniach I (1974) Carcinogenic effect of 100, 200, 250 and 500 rad X-rays on the rat thyroid gland. Br J Cancer 30:487–495

Dunn TB (1975) The unseen fight against cancer: experimental cancer research: its importance to human cancer. Bates, Charlotte, p 111

Ekpechi OL, Dimitriadou A, Fraser R (1966) Goitrogenic activity of cassava (a staple Nigerian food). Nature 210:1137–1138

Ermans AM (1978) Disorders of iodine deficiency: endemic goiter. In: Werner SC, Ingbar SH (eds) The thyroid. Harper and Row, New York, pp 537–553

Escobar del Rey F, Morreale de Escobar (1961) The effect of propylthiouracil, methylthiouracil and thiouracil on the peripheral metabolism of l-thyronine in thyroidectomised,l-thyronine maintained rats. Endocrinology 69:456–465

Fagin JA, Matsuo K, Karmarkar A, Chen DL, Tang SH, Koeffler HP (1993) High prevalence of mutations of the p53 gene in poorly differentiated human thyroid carcinomas. J Clin Invest 91:179–184

Falconer IR, Roitt IM, Seamark RF, Torrigiani G (1965) Studies of the congenitally goitrous sheep. Iodoproteins of the goitre. Biochem J 117:417–424

Flack JD, Hakansson S, Jeffery DJ, Kelvin AS, Maile PA, McCurrdo AS, Perkins CI (1989) Investigation of the effects of diproteverine on the thyroid of the rat. Hum Toxicol 8:411

Gaitan E (1988) Goitrogens. Ballieres Clin Endocrinol Metab 2:683–702

Gartner R, Greil W (1986) The mitogenic activity of IGF1, insulin and EGF on isolated porcine thyroid follicles under negative control of TSH and cAMP. Ann Endocrinol 47:66A

Gmelin R, Virtanen AI (1960) The enzymatic formation of thiocyanate (SCN) from a precursor in Brassica species. Acta Chem Scand 14:507

Greer MA, Studer H, Kendall JW (1967) Studies on the pathogenesis of colloid goiter. Endocrinology 81:623–632

Griesbach WE, Kennedy TH, Purves HD (1945) Studies on experimental goitre VI: thyroid adenomata in rats on Brassica seed diet. Br J Exp Pathol 26:18–24

Grubeck-Loewenstein B, Buchan G, Sadeghi R, Kissonerghis M, Londei M, Turner M, Pirich, Roka R, Niederle B, Kassal H, Waldhausal W, Feldman M (1989) Transforming growth factor β regulates thyroid growth. J. Clin. Invest. 83:764–770

Heldin NE, Westermark B (1991) The molecular biology of the human anaplastic thyroid carcinoma cell. Thyroidol Clin Exp 3:127–131

Hiasa Y, Ohshima M, Kiathori Y, Yuasa T, Fujita T, Iwata C (1982) Promoting effects of 3, amino -1,2,4-triazole on the development of thyroid tumors in rats treated with N-bis (2-hydroxypropyl) nitrosamine. Carcinogenesis 3:381–384

Hill RN, Erdreich LS, Paynter OE, Roberts PA, Rosenthal SL, Wilkinson CF (1989) Thyroid follicular cell carcinogenesis. Fund Appl Toxicol 12:629–697

Jemec B (1977) Studies on the goitrogenic and tumorigenic effect of two goitrogens. Cancer 40:2188–2202

Jemec B (1980) Studies on the goitrogenic and tumourigenic effect of two goitrogens in combination with hypophysectomy or thyroid hormone treatment. Cancer 45:2138–2148

Kohrle J, Hesch RD, Leonard JL (1991) Intracellular pathways of iodothyronine metabolism. In: Braverman LE, Utiger RD (eds) The thyroid. Lipincott, New York, pp 144–189

Lazarus JH (1986) Endocrine and metabolic effects of lithium. Plenum, New York

Ledent C, Dumont JE, Vassart G, Parmentier M (1991) Thyroid adenocarcinomas secondary to tissue-specific expression of simian virus-40 large T antigen in transgenic mice. Endocrinology 129:1391–1401

Ledent C, Dumont JE, Vassart G, Parmentier M (1992) Thyroid expression of an A2 adenosine receptor transgene induces thyroid hyperplasia and hyperthyroidism. EMBO J 11:537–542

Lemoine NR, Mayall ES, Williams ED, Thurston V, Wynford-Thomas D (1988) Agent-specific ras oncogene activation in rat thyroid. Oncogene 3: 541–544

Lemoine NR, Mayall ES, Wyllie FS, Williams ED, Goyns M, Stringer B, Wynford-Thomas D (1989) High frequency of ras oncogene activation in all stages of human thyroid tumorigenesis. Oncogene 4:159–164

Maciel RMB, Moses AC, Villone G, Tramontano D, Ingbar SH (1988) Demonstration of the production and physiological role of insulin-like growth factor II in rat thyroid follicular cells in culture. J Clin Invest 82:1546–1553

Maloof F, Dobyns B, Vickery AL (1952) The effects of various doses of radioactive iodine on the function and structure of the thyroid of the rat. Endocrinology 50:612–638

Many MC, Denef JP, Gathy P, Haumont S (1983) Morphological and functional changes during thyroid hyperplasia and involution in C3H mice: evidence for folliculogenesis during involution. Endocrinology 112:1292–1302

Matovinovic J, Nishiyama RH, Poissant G (1970) Transplantable thyroid tumours in the rat: development of normal appearing thyroid follicles in the differentiated tumors, and development of differentiated tumors from iodine-deficient, thyroxine involuted goiters. Cancer Res 30:504–514

McClain RM (1989) The significance of microsomal enzyme induction and altered thyroid function in rats: implications for thyroid gland neoplasia. Toxicol Pathol 17:294–306

Mohr U, Reznik G, Pour P (1977) Carcinogenic effects of diisopanolnitro-samine in Sprague-Dawley rats. JNCI 58:361–366

Mulligan LM, Kwok JBJ, Healey CS, Elsdon MJ, Eng C, Gardner E, Love DR, Mole SE, Moore JK, Papi L, Ponder MA, Telenius H, Tunnacliffe A, Ponder BAJ (1993) Germ-line mutations of the ret proto-oncogene in multiple endocrine neoplasia type 2A. Nature 363:458–460

Murthy ASK (1980) Morphology of the neoplasms of the thyroid gland in Fischer 344 rats treated with 4,4'-methylene-bis-(N,N'-dimethyl)-benzylamine. Toxicol Lett 6:391–397

Nadler NJ, Mandavia M, Goldberg M (1970) The effect of hypophysectomy on the experimental production of rat thyroid neoplasia. Cancer Res 30: 1909–1911

Nagataki S, Ingbar SH (1986) Autoregulation: effects of iodide. In: Ingbar SH, Braverman LE (eds) The thyroid. Lippincott, Philadelphia, pp 319–330

Nakamura T, Yana I, Kobayashi T, Shin E, Karakawa K, Fujita S, Miya A, Mori T, Nishisho I, Takai S (1993) p53 gene mutations associated with anaplastic transformation of human thyroid carcinomas. Jpn J Cancer Res 83:1293–1298

Nichols CW, Lindsay S, Sheline GE, Chaikoff IL (1965) Induction of neoplasms in rat thyroid glands by X-irradiation of a single lobe. Arch Pathol 80:177–183

Northcutt RC, Steil JN, Hollifield JW, Stant EG (1969) The influence of cholestyramine on thyroxine absorption. JAMA 208:1857–1861

Paynter OE, Burin GJ, Jaeger RB, Gregorio CA (1988) Goitrogens and thyroid follicular cell neoplasia: evidence for a threshold process. Regul Toxicol Pharmacol 8:102–119

Pisarev MA, Chazenbalk GD, Velecchi RM et al. (1986) Action of iodinated derivatives of arachiodonic acid on thyroid growth and cyclic AMP content: possible role in the autoregulatory mechanism. Ann Endocrinol 47:121A

Refetoff S, Larsen PR (1989) Transport, cellular uptake and metabolism of thyroid hormones. In: De Groot L (ed) Endocrinology, vol I. Saunders, Philadelphia, pp 541–561

Roger PP, Dumont JE (1984) Factors controlling proliferation and differentiation of canine thyroid cells cultured in reduced serum conditions: effects of thyrotropin, cyclic AMP and growth factors. Mol Cell Endocrinol 36:79–93

Roger PP, Servais P, Dumont JE (1983) Stimulation by thyrotropin and cyclic AMP of the proliferation of quiescent canine thyroid cells cultured in a defined medium containing insulin. FEBS Lett 157:323–329

Salvatore G, Stanbury JB, Rall JE (1980) Inherited defects of thyroid hormone biosynthesis. In: De Visscher M (ed) Comprehensive endocrinology: the thyroid gland. Raven, New York, pp 443–487

Saunders JE, Eigenburg DA, Bracht LE, Wang WR, van Zweiten MJ (1988) Thyroid and liver trophic changes in rats secondary to microsomal enzyme induction caused by an experimental leukotriene antagonist (L-649,923). Toxicol Appl Pharmacol 5:378–387

Schaller RT, Stevenson JK (1966) Development of carcinoma of the thyroid in iodine-deficient rats. Cancer 19:1063–1080

Smith P, Wynford-Thomas D, Stringer BMJ, Williams ED (1986) Growth factor control of rat thyroid follicular cell proliferation. Endocrinology 119:1439–1445

Stanbury JB, Chapman EM (1960) Congenital hypothyroidism with goitre: absence of an iodide concentrating mechanism. Lancet i:1162–1165

Stanbury JB, Hedge AN (1950) A study of a family of goitrous cretins. J Clin Endocrinol 10:1741–1758

Stanbury JB, Morris (1958) Deiodination of di-iodotyrosine by cell-free systems. J Biol Chem 233:106–108

Story DL, Cardona RA, Lengen MR (1993) Effect of dietary PCNB on circulating levels of T3 and T4 and TSH in rats. Toxicologist 13:1446

Taurog A (1976) The mechanism of action of the thiourylene antithyroid drugs. Endocrinology 98:1031–1046

Thilly CH, Swennen B, Bourdoux P, Ntambue K, Moreno-Royes R, Gillies J, Vanderpas JB (1993) The epidemiology of iodine-deficiency disorders in relation to goitrogenic factors and thyroid stimulting hormone regulation. Am J Clin Nutr 57:267S–270S

Thomas JA, Bell JU (1982) Endocrine toxicology. In: Hayes AW (ed) Principles and methods in toxicology. Raven, New York, pp 487–496

Thomas GA, Williams ED (1991) Evidence for and possible mechanisms of a non-genotoxic carcinogenesis in the rodent thyroid. Mutat Res 248:357–370

Thomas GA, Williams ED (1992) Production of thyroid tumours in mice by demethylating agents. Carcinogenesis 13:1039–1042

Thomas GA, Williams ED (1994) Age related changes in structure and function of the thyroid follicular cell. In: Capen CC, Mohr U (eds) Pathology of the aging rat, vol 2. ILSI, Washington, pp 269–283

Thomas GA, Williams D, Williams ED (1988) The demonstration of tissue clonality by X-linked enzyme histochemistry. J Pathol 155: 101–108

Thomas GA, Williams D, Williams ED (1989) The clonal origin of thyroid nodules and adenomas. Am J. Pathol 134:141–147

Thomas GA, Williams D, Williams ED (1991) Reversibility of the malignant phenotype in monoclonal thyroid tumours in the mouse. Br J Cancer 63:213–216

Thomas GA, Davies HG, Williams ED (1994) Expression of IGF1 in the normal and short-term goitrogen treated mouse thyroid. J Pathol (in press)

Todd GC (1986) Induction and reversibility of thyroid proliferative changes in rats given an antithyroid compound. Vet Pathol 23:110–117

Tsuda H, Fukunshima S, Imaida K, Kurata Y, Ito N (1983) Organ-specific promoting effect of phenobarbital and saccharin in induction of thyroid, liver and bladder tumors in rats after initiation with N-nitrosomethylurea. Cancer Res 43:3292–3296

Van Jaarsveld P, van der Walt B, Theron CN 91972) Afrikander cattle congenital goiter: purification and partial identification of the complex iodoprotein pattern. Endocrinology 91:470–482

Van Sande J, Dumont JE (1973) Effects of thyrotropin, prostaglandin E1 and iodide on cyclic 3'5'AMP concentration in dog thyroid slices. Biochim Biophys Acta 313:320

Vickery AL (1981) The diagnosis of malignancy in dyshormonogenetic goitre. Clin Endocrinol Metab 10:317–335

Westermark B, Karlson FA, Walinder D (1979) Thyrotropin is not a growth factor for human thyroid cells in culture. Proc Natl Acad Sci USA 76:2022-2026

Williams DW, Wynford-Thomas D, Williams ED (1987) Control of human thyroid follicular cell proliferation in suspension and monolayer culture. Mol Cell Endocrinol 51: 33–40

Williams DW, Williams ED, Wynford-Thomas D (1988) Loss of dependence on IGF-1 for proliferation of human thyroid adenoma cells. Br J Cancer 57:535–539

Williams DW, Williams ED, Wynford-Thomas D (1989) Evidence for autocrine production of IGF-1 in human thyroid adenomas. Mol Cell Endocrinol 61:139–143

Williams ED, Doniach I (1963) Thyroid autografts in hypophysectomised and thyroxine treated rats. J Endocrinol 26:479–488

Williams ED, Doniach I, Bjarnason O, Michie W (1977) Thyroid cancer in an iodide rich area. Cancer 39:215–222

Wolff J, Maurey J (1961) Thyroidal iodide transport II. Comparison with non-thyroid iodide concentrating tissues. Biochim Biophys Acta 47:467-474

Wollman SH, Breitman TR (1970) Changes in DNA and weight of thyroid glands during hyperplasia and involution. Endocrinology 86:322–327

Wynford-Thomas D, Stringer BMJ, Williams ED (1982a) Goitrogen induced thyroid growth in the rat: a quantitative morphometric study. J Endocrinol 94:131–140

Wynford-Thomas D, Stringer BMJ, Williams ED (1982b) Dissociation of growth and function in the rat thyroid during prolonged goitrogen administration. Acta Endocrinol (Copenh) 101:210–216

Wynford-Thomas D, Stringer BMJ, Williams ED (1982c) Desensitisation of rat thyroid to the growth stimulating action of TSH during prolonged goitrogen administration. Acta Endocrinol (Copenh) 101:562–569

Wynford-Thomas D, Stringer BMJ, Harach HR, Williams ED (1985) Mitotic response in goitrous and normal rat thyroid: implications for thyroid growth control. Cell Tissue Kinet 18:467–473

Wyngaarden JB, Stanbury JB, Rapp B (1953) The effect of iodide, perchlorate, thiocyanate and nitrate administration upon the iodide concentrating mechanism of the rat thyroid. Endocrinology 52:568–574

6 Nongenotoxic Carcinogenesis in the Liver

R. Schulte-Hermann, W. Bursch, B. Grasl-Kraupp, W. Huber, and W. Parzefall

6.1 Introduction

Numerous compounds have been found to produce tumors in rodent liver but have shown no evidence of genotoxic activity. These nongenotoxic compounds comprise a diverse group of important drugs and environmental pollutants. Examples are shown in Table 1 (Schulte-Hermann 1985; Schulte-Hermann et al. 1990b). Some agents, for example, certain steroids, have been associated with tumor formation in *human* liver (Schulte-Hermann 1985; Schulte-Hermann et al. 1990a; Tao 1991), and ethanol can be classified as a human nongenotoxic carcinogen, too. Therefore, hepatocarcinogenesis induced by nongenotoxic agents is not only a problem in rodents, and assessment of human health risks is urgently required. For this purpose we need to know the relevant

Table 1. Some nongenotoxic hepatocarcinogens and their acute hepatic effects

Prototype	Induction of Growth	Monooxygenase	Other
Phenobarbitol, DDT, HCH, some PCB	+	P450 II B,C isoforms	+
TCDD, some PCBs	+	P450 I A	+
Estradiol esters, ethinylestradiol	+	–	Clotting factors, angiotensinogen, etc.
Clofibrate, diethyl-hexylphthalate, nafenopin	+	P450 IV	Some peroxisomal enzymes
Thioacetamide, thiobenzamide	+	–	Phase II enzymes
Ethanol	+	P450 II E1	+
CCl4	+	–	

DDT, dichlorodiphenyl-trichloroethane; HCH, hexachlorocyclohexane; PCB, poly-chlorinated biphenyls; TCDD, tetrachlorodibenzodioxin.

biological effects of these agents in rodent and human liver as well as the mechanisms underlying carcinogenesis.

6.2 Effects of Nongenotoxic Carcinogens on Normal Liver

Virtually all nongenotoxic liver carcinogens induce growth in their target organ (Table 1). The type of growth can be classified either as regenerative or adaptive (or a combination of both). In the former case a cytotoxic effect leading to cell injury and death is the primary event, and subsequently enhanced DNA synthesis and mitosis serve to replace dying cells. CCl4 can be considered as an example. It seems important to note that this type of effect is associated with steadily enhanced cell replication if cytotoxic injury persists, such as during chronic treatment.

The other, adaptive type of growth is accompanied by functional changes such as increases in activities of drug-metabolizing enzymes, of enzymes involved in fatty acid metabolism, or of other proteins. In other words, activation of gene programs or pleiotropic responses are induced by agents inducing adaptive growth. In some cases (tetrachlordiben-zodioxin, TCDD, estrogens, peroxisome proliferators) these responses appear to be mediated by specific receptors. In general, nongenotoxic compounds stimulating adaptive growth in the liver seem to act like trophic hormones in their respective target organs.

Typically, the growth response occurs through an initial burst of enhanced DNA synthesis followed by enhanced mitosis (at least in rat liver). It is important that enhanced proliferation of cells ceases after a few days even if treatment is continued. Obviously, the agents produce a new steady state of cell number in the liver at an enhanced level; this hyperplasia is sustained as long as treatment is performed. In the state of hyperplasia, liver cells are resistant to further stimulation of DNA synthesis. Obviously, an effective feedback mechanism prevents excessive cell multiplication in the liver even if the growth stimulatory signals are steadily present due to continuous compound treatment.

When treatment is stopped, rapid regression of hyperplasia may occur, depending on the compound. This regression in the liver is due to a steep increase in the rate of active cell death, frequently of the apoptotic type (Bursch et al. 1984). Active cell death is a process whereby organisms can actively eliminate excessive ("unwanted") or damaged cells. It functions complementary to mitosis in order to maintain an adequate size and cell number in tissues (Wyllie et al. 1980). Active cell death can be inhibited by tissue-specific growth factors and mitogens. Our observations led to the conclusion that, in the state of sustained hyperplasia, some of the liver cells depend on the continuous application of a liver mitogen, otherwise they will die. In other words during sustained hyperplasia there is no active cell proliferation but some "growth pressure" appears to be active. This growth pressure is important for carcinogenesis by agents stimulating adaptive growth.

A third type of nongenotoxic liver carcinogens is exemplified by agents such as methapyrilene, thioacetamide, or ethanol, which have some enzyme-inducing (and adaptive) potential and at the same time are cytotoxic (Table 1). They may act through either adaptive or regenerative mechanisms or by a combination of the two.

6.3 Effects of Nongenotoxic Carcinogens
on Stages of Hepatocarcinogenesis

It is now generally accepted that cancer formation in the liver and other organs occurs through a sequence of cellular intermediates. At least three different stages have been defined, namely, initiation, promotion, and progression (Schulte-Hermann 1985; Pitot and Sirica 1980; Farber 1991). In the liver, early intermediates exhibit certain phenotypic alterations detectable by histological, histochemical, or immunological stains. Studies with phenobarbital first performed by Peraino, and later by numerous other groups, have shown that most of the nongenotoxic liver carcinogens listed in Table 1 can accelerate tumor formation from lesions initiated by application of genotoxic carcinogens, indicating that they are liver tumor promoters (Schulte-Hermann 1985). In some studies ethanol was also found to be a tumor promoter (Seitz et al. 1989). Phenobarbital and many other agents exert their promoting effect by stimulation of growth of putative preneoplastic foci made visible by staining for γ-glutamyl transferase (GGT), glutathione-S-transferase (GST-P), or other markers (Schulte-Hermann 1985; Pitot and Sirica 1980; Farber 1991; Schulte-Hermann et al. 1990b).

Peroxisome proliferators as a rule did not stimulate growth of these GGT or GST-P positive foci and did not appear to promote hepatocarcinogenesis in a number of early studies (Numoto et al. 1985; Williams et al. 1987; Popp et al. 1985). However, more recently we and others have shown that this class of compounds promotes growth of a special type of putative preneoplastic liver foci which appear phenotypically and biologically different from the GGT/GST-P-positive foci (Kraupp-Grasl et al. 1990, 1991). Due to their appearance in sections stained with H&E we introduced the term "weakly basophilic" for these lesions (Table 2).

The reasons for the divergent response of foci classes to phenobarbital type promoters or peroxisome proliferators are unknown. It may be relevant in this context that GGT/GST-P-positive foci frequently overexpress those drug-metabolizing enzymes which can be induced in normal liver by phenobarbital (Schulte-Hermann et al. 1986). In contrast, weakly basophilic foci show no or very little expression of drug-metabolizing enzymes but tend to overexpress peroxisomal enzymes (Grasl-Kraupp et al. 1993). Associated with these functional changes the two foci subtypes also overrespond to the growth stimulatory effects

Table 2. Types of phenotypically altred liver foci and some of their characteristics

	Eosinophilic/ clear cell foci	Weakly baso-philic foci	Tigroid foci
Promoter	PB	NAF	?
Cell turnover:			
Replication	+	+	+
Cell death (apoptosis)	+	+	+
GGT, GST-P	+	–	–
Functional alterations:			
Glycogen storage	+	–	–
Drug metabolism	+	–	?
Peroxisomal enzymes	–	(+)	–

+, Stronger than in normal liver; –, less than in normal liver or negative; PB, pheno-barbital; NAF, nafenopin; GGT, γ-glutamyl transferase; glutathione-S-transferase (GST-P).

of the nongenotoxic carcinogens/tumor promoters. Therefore, we hypo-thesize that an important regulatory defect in putative preneoplastic foci may render cells overresponsive to the pleiotropic effects of promoters. Consistent with this hypothesis are recent data by Jirtle suggesting that foci cells, in contrast to normal liver cells, do not exhibit resistance to repeated mitogenic signals (Jirtl and Meyer 1991).

Studies on the mechanism of growth of preneoplastic foci revealed that these lesions exhibit severalfold higher rates of cell proliferation than normal liver cells; nevertheless, the growth rate without promoter treatment was very low. It was then found that apoptotic activity was also higher in foci and largely counterbalanced the increase of cell proliferation (Bursch et al. 1984; Schulte-Hermann et al. 1990b). Apparently apoptosis is selectively increased in preneoplastic foci. How this selectivity is achieved is unknown. Phenobarbital and other promoters can inhibit apoptosis and thereby accelerate foci growth (Bursch et al. 1984; Schulte-Hermann et al. 1990b). This observation supports a specific growth stimulatory effect of promoters on preneoplastic cells. In conclusion, enhanced apoptosis in preneoplastic cells seems to be one of the defense lines against cancer formation and is disturbed by tumor promoters.

6.4 Role of "Spontaneous" Initiation for the Mechanism of Action of Nongenotoxic Carcinogens

As outlined above, most or all nongenotoxic hepatocarcinogens are tumor promoters. How can tumor promoters produce tumors if given without initiating pretreatment according to the present routine testing procedures for carcinogenicity? Two basically different possibilities must be envisaged: (1) The agents may have direct or indirect initiating activity not readily detected by current genotoxicity tests; (2) initiated cells may form independently of the compound ("spontaneously"), and promoters may produce tumors by promoting development of these spontaneously initiated cells. We have specifically studied possibility 2 in recent years using the peroxisome proliferator nafenopin as a model compound, and some interesting findings will be reported below.

Preneoplastic foci of seemingly "spontaneous" origin have previously been observed in rat liver by our group as well as by others (Ward 1983; Schulte-Hermann et al. 1983). These spontaneous foci are rarely found in young rats but frequently in old rats. In order to test whether nafenopin would be able to promote development of such spontaneous preneoplastic foci to carcinomas, we treated either young or old rats for identical periods (1 year) with identical doses of nafenopin. At the beginning of treatment the young rats contained virtually no detectable foci in the livers, while the old rats showed numerous foci. After 1 year of treatment very few liver tumors were found in the animals young at the onset of experimentation but numerous adenomas and carcinomas were present in the group of old animals. Untreated rats, either young or old, did not exhibit liver tumors, with the exception of a single lesion in an old animal (Kraupp-Grasl et al. 1991).

This finding is clearly consistent with the hypothesis that carcinogenesis by peroxisome proliferators require the presence of "spontaneous" preneoplastic lesions. On the other hand, and in view of the hypothesis that peroxisome proliferators should produce oxidative stress to hepatocytes and thereby be carcinogenic (Reddy and Lalwai 1983), it might be argued that old rats may be less able to handle damage by oxidative stress. Therefore we studied various indicators of oxidative stress in the livers of young and old rats. No significant differences between the two groups could be found (Huber et al. 1991). Thus the

Table 3. Effect of food restriction to 60% of the control amount for 3 months on foci growth, as determind by DNA synthesis, apoptosis and foci volume and on subsequent tumor development in the liver

	Ad libitum	restricted
DNA synthesis (%)	0.538 ± 0.660^a	0.088 ± 0.105^a
Apoptosis (%)	0.107 ± 0.126^a	0.33 ± 0.32^a
Volume/liver	$12.7 \pm 6.1 \text{ mm}^3$	$1.9 \pm 1.7 \text{ mm}^3$
Tumors/liver	45.5 ± 25.3	21.3 ± 9.2

[a]Averages and standard deviations.

only difference between young and old rat livers detected so far was the presence or absence of putative preneoplastic cells. It may be of interest to add that there was no evidence of enhanced lipid peroxidation as indicated by malone dialdehyde formation following nafenopin treatment (Huber et al. 1991). Even after a massive dietary load with lipids, no enhanced signs of lipid peroxidation were observed in the liver of nafenopin pretreated rats (Huber et al., unpublished). These observations do not support the oxidative stress hypothesis of peroxisome proliferator carcinogenesis.

If it were true that peroxisome proliferators are carcinogenic only by promotion of spontaneously initiated cells, then elimination of such cells should eliminate the carcinogenic activity of these agents. Is it possible to eliminate initiated cells from the liver? As mentioned above, putative preneoplastic cells in liver foci show relatively high rates of apoptosis. This led to the hypothesis, based on mathematical modeling of carcinogenesis, that most initiated cells in the liver are extinguished before they develop into foci or tumors (Luebeck et al. 1991). Some biological experiments also support this hypothesis. Our approach to eliminate spontaneously initiated cells was based on the following rationale: We knew from previous experiments that food restriction increases the rate of apoptosis in normal liver (Schulte-Hermann et al. 1988). We also knew that preneoplastic cells are more sensitive to signals inducing apoptosis than normal liver cells (see above). Consequently, we postulated that a period of fasting should preferentially eliminate initiated cells.

To test this hypothesis we subjected 1-year-old rats containing numerous putative preneoplastic foci of spontaneous origin to food

Table 4. Effect of 17 months of nafenopin treatment and 1 month of withdrawal on preneoplastic and neoplastic lesions in rat liver

	+ Nafenopin	Nafenopin withdrawn for 34 days
Tumors/liver	45.5 ± 25.0	6.3 ± 3.7
Weakly basophilic foci (mm^3/liver)	71.6 ± 55.1	3.5 ± 3.5

reduction by 40% for a period of 3 months. As shown in Table 3, this resulted in a decrease of DNA synthesis and an increase of apoptosis in foci. At the end of the 3-month restriction period foci volume and number had decreased to 10%–20%. Subsequently, animals were fed ad libitum and a subgroup was placed on a diet with nafenopin. After 1.5 years of treatment the tumor yield in the group with restricted diet, which had less foci at the beginning of nafenopin treatment, was lower by approximately 50% (Table 3). This result confirms the hypothesis that the carcinogenicity of nafenopin depends on the presence of spontaneously initiated cells. Furthermore, this finding is of interest from a more general point of view because it provides a mechanism to explain the protection from cancer development by reduced feeding, which is well known to occur in animals and humans.

6.5 Reversibility of Tumors Produced by Nafenopin Treatment

When nafenopin treatment was stopped at a stage when multiple tumors were present in the liver, the majority of adenomas and even of carcinomas disappeared within 5 weeks (Table 4). During this time there was a complete cessation of DNA synthesis in liver lesions, and apoptosis and lytic cell death increased dramatically. These findings show that cells in nafenopin-induced tumors even at the carcinoma stage still exhibit dependence on the growth stimulatory signals exerted by the peroxisome proliferator. This observation is consistent with the postulated hormonelike mechanism of action of nongenotoxic carcinogens. The persistance of a few tumors after 5 weeks of nafenopin withdrawal can easily be explained by spontaneous progression of lesions towards promoter-independent growth. Alternatively, 5 weeks might be too short to allow for complete regression of *all* tumors. In conclusion, the rapid and extensive regression of liver tumors produced by nafenopin does not support an initiating or complete carcinogenic effect of this agent.

6.6 Implications for Risk Assessment

In conclusion, our results suggest that the hepatocarcinogenic effect of the model compound nafenopin is due to the promotion of initiated cells formed spontaneously in the liver; these cells may specifically overexpress the pleiotropic response to the promoting agent. The findings do not completely exclude the possibility that nafenopin has initiating activity in addition to its promoting potential, but there is no need or evidence to classify nafenopin as a complete carcinogen. Similar conclusions are probably valid for most other nongenotoxic hepatocarcinogens.

If the hypothetical mechanism of hepatocarcinogenic action of nongenotoxic carcinogens discussed above is correct, two important questions are raised: Firstly, does human liver contain "spontaneously" initiated cells with properties similar to those described in rat liver? In this case, application of a tumor promoter could lead to cancer formation. This question cannot be answered yet but the formation of adenomas in human liver by tumor-promoting steroids suggests that such types of lesions may well occur, albeit rarely. Secondly, does human liver show the same type of pleiotropic response to nongenotoxic carcinogens as rat liver, and if so, at what dose? For peroxisome proliferators present evidence suggests that human hepatocytes show little, if any, peroxisome proliferation or DNA synthesis in response to these agents (Blaaboer et al. 1990; Bichet et al. 1990; Parzefall et al. 1991). In contrast, phenobarbital and similar agents have frequently been shown to induce multiplication of smooth endoplasmic reticulum and drug-metabolizing enzymes in human liver; however, induction of DNA synthesis or liver growth has not been clearly demonstrated. Furthermore, epidemiological studies on rather large groups of patients treated for decades with high doses of phenobarbital have not revealed a tumorigenic risk (Clemmesen and Hjalgrim-Jensen 1978). Therapeutic doses of phenobarbital in humans may be even higher than the no-observed effect level for tumor promotion in rats. Thus, pleiotropic responses may partially occur in human liver, depending on the agent, but overall sensitivity to promotion by phenobarbital or peroxisome proliferators seems to be lower than in rat liver.

References

Blaaboer BJ, VanHolstein CWM, Bleumink R, Mennes WC, VanPelt FNAM, Yap SH, VanPelt JF, VanIersel AAJ, Timmermann A, Schmid BP (1990) The effect of beclobric acid and clofibric acid on peroxisomal β-oxidation and peroxisome proliferation in primary cultures of rat, monkey and human hepatocytes. Biochem Pharmacol 40:521–528

Bichet N, Cahard D, Fabre G, Remandet B, Gouy D, Cano JP (1990) Toxicological studies on a benzofurane derivative. III. Comparison of peroxisome proliferation in rat and human hepatocytes in primary culture. Toxicol Appl Pharmacol 106:509–517

Bursch W, Lauer B, Timmermann-Trosiener I (1984) Controlled death (apoptosis) of normal and putative preneoplastic cells in rat liver following withdrawal of tumor promoters. Carcinogenesis 5:453–458

Clemmesen J, Hjalgrim-Jensen S (1978) Is phenobarbital carcinogenic? A follow-up of 8078 epileptics. Ecotoxicol Environ Safety 1:457–470

Farber E (1991) Hepatocyte proliferation in stepwise development of experimental liver cell cancer. Dig Dis Sci 36:973–978

Grasl-Kraupp B, Waldhör T, Huber W, Schulte-Hermann R (1993) Glutathione-transferase-isoenzyme patterns in different subtypes of enzyme altered rat liver foci treated with the peroxisome proliferator nafenopin or with phenobarbital. Carcinogenesis 14:2407–2412

Huber W, Kraupp-Grasl B, Esterbauer H, Schulte-Hermann R (1991) Role of oxidative stress in age dependent hepatocarcinogenesis by the peroxisome proliferator nafenopin in the rat. Cancer Res 51:1789–1792

Jirtle RL, Meyer SA (1991) Liver tumor promotion: effect of phenobarbital on EGF and protein kinase C signal transduction and transforming growth factor-β1 expression. Dig Dis Sci 5:659–668

Kraupp-Grasl B, Huber W, Putz B, Gerbracht U, Schulte-Hermann R (1990) Tumor promotion by the peroxisome proliferator nafenopin involving a specific subtype of altered foci in rat liver. Cancer Res 50:3701–3708

Kraupp-Grasl B, Huber W, Taper H, Schulte-Hermann R (1991) Increased susceptibility of aged rats to hepatocarcinogenesis by the peroxisome proliferator nafenopin and the possible involvement of altered liver foci occurring spontaneously. Cancer Res 51:666–671

Luebeck EG, Moolgavkar SH, Buchmann A, Schwarz M (1991) Effects of polychlorinated biphenyls in rat liver: quantitative analysis of enzyme-altered foci. Toxicol Appl Pharmacol 111:469–484

Numoto S, Mori H, Furuya K, Levine WC, Williams GM (1985) Absence of a promoting or sequential syncarcinogenic effect in rat liver by the carcinogenic hyperlipidemic drug nafenopin given after N-2-fluorenylacetamide. Toxicol Appl Phamacol 77:76–85

Parzefall W, Erber E, Sedivy R, Schulte-Hermann R (1991) Testing for indication of DNA synthesis in human hepatocyte primary cultures by rat liver tumor promoters. Cancer Res 51:1143–1147

Pitot HC, Sirica AE (1980) The stages of initiation and promotion in hepatocarcinogenesis. Biochim Biophys Acta 605:191–215

Popp JA, Garvey LK, Hamm TE Jr, Swenberg JA (1985) Lack of hepatic promotional activity by the peroxisomal proliferating hepatocarcinogen di(2-ethylhexyl)phthalate. Carcinogenesis 6:141–144

Reddy JK, Lalwai ND (1983) Carcinogenesis by hepatic peroxisome proliferators: evaluation of the risk of hypolipidemic drugs and industrial plasticisers to humans. Crit Rev Toxicol 12:1–58

Schulte-Hermann R (1985) Tumor promotion in the liver. Arch Toxicol 57:147–158

Schulte-Hermann R, Timmermann-Trosiener I, Schuppler J (1983) Promotion of spontaneous preneoplastic cells in rat liver as a possible explanation of tumor production by non-mutagenic compounds. Cancer Res 43:839–844

Schulte-Hermann R, Timmermann-Trosiener I, Schuppler J (1986) Facilitated expression of adaptive responses to phenobarbital in putative pre-stages of liver cancer. Carcinogenesis 10:1651–1655

Schulte-Hermann R, Bursch W, Fesus L, Kraupp B (1988) Cell death by apoptosis in normal, preneoplastic and neoplastic tissue. In: Feo F, Pani P, Columbano A, Garcea R (eds) Chemical carcinogenesis: models and mechanisms. Plenum, New York, pp 263–274

Schulte-Hermann R, Parzefall W, Bursch W, Ochs H, Kraupp B (1990a) Nicht-gentoxische Kanzerogene. In: Grosdanoff P, Kraupp O, Schütz W, Schulte-Hermann R (eds) Toxikologische und klinisch-pharmakologische Prüfungen. De Gruyter, Berlin, pp 239–256

Schulte-Hermann R, Timmermann-Trosiener I, Barthel G, Bursch W (1990b) DNA synthesis, apoptosis, and phenotypic expression as determinants of growth of altered foci in rat liver during phenobarbital promotion. Cancer Res 50:5127–5135

Seitz HK, Simanowski, Hoerner M, Kommerell B (1989) Alcohol and liver carcinoma. In: Bannasch B, Keppler D, Weber G (eds) Liver cell carcinoma. Kluwer Academic, Dordrecht, pp 227–241

Tao L-C (1991) Oral contraceptive-associated liver cell adenoma and hepatocellular carcinoma. Cancer 68:341–347

Ward JM (1983) Increased susceptibility of livers of aged F344/NCr rats to the effects of phenobarbital on the incidence, morphology, and histochemistry of hepatocellular foci and neoplasms. J Natl Cancer Inst 71:815–822

Williams GM, Maruyama H, Tanaka T (1987) Lack of rapid initiating, promoting or sequential syncarcinogenic effects of di(2-ethylhexyl)phthalate in rat liver carcinogenesis. Carcinogenesis 8:875–880

Wyllie AH, Kerr JFR, Currie AR (1980) Cell death: the significance of apoptosis. Int Rev Cytol 68:251–300

7 Compensatory Cell Proliferation, Mitogen-Induced Liver Growth and Hepatocarcinogenesis in the Rat

G. M. Ledda-Columbano and A. Columbano

7.1 Introduction

The liver is an organ constituted largely of epithelial cells, with hepatocytes representing approximately 90% of the cell mass in both rats and mice. In adult organisms, hepatocytes have a long life span, ranging from 200 to 400 days or more. However, it is well known that liver cells divide in response to hepatic cell loss such as that occurring after surgical partial hepatectomy (PH) or cell death induced by chemical hepatotoxins such as carbon tetrachloride (CCl_4) and others. What mechanism is responsible for triggering the proliferative response of the liver is not known. However, since in all of these cases hepatic growth is a compensatory response to decreased liver mass, a fall in the concentration of an inhibiting factor due to cellular loss has long been considered to be the cause of the proliferative response of liver cells.

This type of mechanism, however, cannot apply to conditions under which DNA synthesis and replication of liver cells occur in the absence of cell loss. This condition, defined as additive liver growth, augmentative hyperplasia, or direct hyperplasia, may be achieved following treatment with chemicals (so called primary mitogens) that cause minimal or no hepatocyte necrosis.

While several studies have described the sequence of events taking place during compensatory regeneration following PH, very few have been undertaken to characterize the liver growth induced by primary mitogens at a biochemical and molecular level.

What are the basic differences between compensatory regeneration and mitogen-induced direct hyperplasia? When cell proliferation occurs by a compensatory mode such as that induced by PH or CCl_4, cell loss is the primary event and cell proliferation is triggered in order to replace the cells that were lost. This is why we have termed this type of cell proliferation compensatory regeneration. It is needed to compensate for the previous cellular loss. If we now look at the proliferative process induced by primary mitogens, the situation is quite different in that DNA synthesis is induced in the absence of any previous cell loss, suggesting that these chemicals somehow interfere with the growth control regulatory mechanisms normally operating in the organ. The consequence of this is that the mitogenic stimulus, which is now the primary event, results in an excess of the liver mass, excess DNA, and presumably an excess of cells. Interestingly, as soon as the mitogenic stimulus is withdrawn, liver mass and DNA content go back to their original values, indicating that this regression process, like the proliferative process seen during regeneration, is also strictly controlled. The rapid elimination observed during the regression phase appears to be due to a particular type of cell death, namely, apoptosis (Wyllie et al. 1980).

7.2 Compensatory Regeneration vs Mitogen-Induced Liver Growth

7.2.1 Growth Factors

The fact that the proliferative response seen following mitogens, unlike compensatory regeneration, is not due to any previous reduction in cell number may be interpreted as suggesting that the type of signals responsible for the triggering of the entry into cell cycle might be different. As far as compensatory regeneration is concerned, there are several growth factors that participate in the regulation of hepatocyte proliferation in vivo. These include hepatocyte growth factor/scatter factor (HGF/SF) (Matsumoto and Nakamura 1991; Michalopoulos 1990), transforming growth factor-α (TGF-α) and transforming growth factor-beta 1 (TGF-β1) (Fausto and Mead 1989). While HGF/SF and TGF-α stimulate, TGF-β1 inhibits proliferation of adult hepatocytes in culture. Although it is evident that liver cell growth in vivo is controlled by an interplay of both positive and negative stimuli provided by growth factors, several studies have suggested that HGF/SF may be the critical hepatotrophic factor for the triggering of liver regeneration. Indeed, HGF/SF mRNA and HGF/SF activities were found to be markedly elevated in the liver or plasma of rats after various insults (Asami et al. 1991; Michalopoulos 1990). More recently, it was proposed that a protein, injurin, appearing in the blood of rats subjected to injury may be involved in the regeneration of the organ through its capacity to increase the synthesis of HGF/SF in distant sites such as kidney or lung (Matsumoto et al. 1992). This hypothesis may explain why HGF/SF levels increase during liver regeneration following injury. However, as already mentioned, there is no cell death and/or loss during mitogen-induced liver growth. Therefore, it was of interest to us to determine whether HGF/SF mRNA expression could be induced also under conditions where there had been no organ injury. To study this problem we used lead nitrate (LN) as a direct liver mitogen. A single dose of this chemical induces DNA synthesis, which is followed by an increase in the mitotic index (Columbano et al. 1983). The result of this process is a doubling of liver mass and DNA content 3 days after treatment. To determine the effect of LN-induced liver growth on the expression of HGF mRNA, Wistar rats were killed 1, 3, 5, 10, 24, and 48 h after treatment. Northern analysis of

HGF/SF hepatic mRNA indicated that liver growth induced by the mitogen LN, unlike liver regeneration induced by PH or CCl4, does not require an increased expression of liver HGF/SF mRNA (Shinozuka et al. 1993). Interestingly, no difference in the levels of expression of TGF-α was apparent between controls and LN-treated rats at any time point examined. These results are in agreement with those of Masuhara et al. (1992), indicating that liver growth induced by the mitogen BR931, unlike compensatory regeneration induced by a choline-deficient diet also fails to induce HGF/SF or TGF-α mRNA in rat liver, suggesting that mitogen-induced liver growth and compensatory regeneration may be induced through different mechanisms. The lack of increase in HGF mRNA does not mean that mitogens lack the property of inducing humoral factors involved in cell proliferation. In fact, serum taken from LN-treated rats is able to stimulate DNA synthesis when added to hepatocyte primary cultures as efficiently as the serum taken from rats subjected to PH (Coni et al. 1992). The same effect was observed when serum isolated from rats treated with other liver mitogens, namely, cyproterone acetate (CPA) and ethylene dibromide (EDB), was added to hepatocytes in culture. Therefore, while humoral factors involved in the triggering of cell proliferation are present in the serum of rats during compensatory regeneration and mitogen-induced liver growth, hepatic induction of the growth factors HGF/SF and TGF-α occurs only in the former proliferative stimulus. It is possible that other growth factors may be more critical during mitogen-induced cell proliferation. Recently, it was shown that in vivo administration of tumor necrosis factor-α (TNF-α) is able to stimulate proliferation of nonparenchymal and parenchymal cells (Beyer et al. 1990; Feingold et al. 1991). In an attempt to determine whether LN-induced liver growth could be mediated by an increased TNF-α mRNA expression, we have studied the expression of this gene following LN treatment. The results indicate that LN treatment is indeed able to stimulate TNF-α mRNA as early as 2 h after treatment, suggesting that cell proliferation induced by LN might be mediated by TNF-α (Shinozuka et al. 1993). However, it remains to be established whether the effect of TNF-α (if any) on hepatocyte proliferation is a direct effect or whether it is due to stimulation by TNF-α of other lymphokines, which may in turn induce hepatocyte proliferation.

	COMPENSATORY REGENERATION		DIRECT HYPERPLASIA			
	PH	CCl$_4$	CPA	NAF	LN	EDB
c-fos	+	+	−	−	±	−
c-jun	+	+	−	−	+	+
c-myc	+	+	−	−	+	+
c-Ha-ras	+	+	+	+	+	+
TS	+	+	+	+	+	+

Fig. 1. Schematic representation of the pattern of expression of some cell-cycle related genes during cell proliferation induced by different proliferative stimuli. *TS*, thymidilate synthase; *PH*, partial hepatectomy; *CPA*, cyproterone acetate; *NAF*, nafenopin; *EDB*, ethylene dibromide

7.2.2 Immediate Early Genes

It is known that DNA synthesis induced in hepatocyte culture by addition of growth factors is preceded by an increased expression of some immediate early genes or primary response genes. A similar expression of these genes also occurs during liver regeneration following PH. Several studies (Goyette et al. 1983; Makino et al. 1984; Morello et al. 1990) have shown that expression of immediate early genes during liver regeneration has three main characteristics: (1) it is specific, in that certain protooncogenes, but not others, are involved (fos, jun, myc, and ras are elevated, while abl, mos, src are not involved); (2) it is sequential, in that mRNA from specific protooncogenes increases at well-defined periods after PH, following the general order fos, jun, myc, p53, ras; (3) it is transient, in that increased transcription for these various genes is found only during a period of few hours.

In order to study whether the differences seen in the expression of growth factors such as HGF/SF and TGF-α could also result in a different pattern of expression of cell cycle-related genes, we have examined the pattern of some immediate early genes during liver regeneration induced by PH or mitogen-induced liver growth induced by LN and other liver mitogens, i.e., EDB, CPA, and nafenopin (NAF). The results indicate that while an increased expression of these genes is always seen during compensatory regeneration, no such increase is observed during liver growth induced by certain mitogens. Of particular interest was the finding that an almost undetectable increase in the expression of c-fos was observed during mitogen-induced liver growth, irrespective of the mitogen used (Coni et al. 1990). It is also of interest to notice that liver growth induced by two mitogens, CPA and NAF, was not associated with increased levels of c-fos, c-jun, and c-myc mRNA, suggesting that under some conditions liver growth in vivo may not necessarily require an increased expression of these immediate early genes (Coni et al. 1993). In contrast, when the expression of genes associated to "S" phase or later phases of the cell cycle was analyzed, no differences between the two proliferative stimuli could be found. Figure 1 schematically illustrates the pattern of some cell-cycle related genes observed during different types of liver cell proliferation.

Liver response to mitogens thus appears to be different from that elicited by regenerative stimuli not only in terms of growth factors but also in the expression of immediate early genes.

7.2.3 Gap Junction Intercellular Communications

At a cellular level, one of the features accompanying the production of proliferative stimuli is the loss of gap junction intercellular communications (GJIC). It is believed that under normal conditions, replicative signal substances are produced periodically within cells and pass into adjacent cells through gap junctions (Beyer et al. 1990; Hertzberg and Johnson 1988). These signals, possibly cyclic nucleotides or ions, would activate cell division if they reached a sufficient and sustained concentration in the signal-producing cell. If the level of these chemical signals is maintained in a reduced concentration by intercellular communication, cell division would not occur in the signal-producing cell. It is now well known that after PH a loss of communication between cells occurs as a consequence of a reduction of GJIC. Dermietzel et al. (1987), Traub et al. (1989), and Neveu et al. (1990) have shown that the appearance of two connexins, connexin 32 (cx32) and connexin 26 (cx26), is reduced after PH, suggesting that these connexins (CXs) might be controlled by the same mechanism. In order to investigate whether the differential effect induced by the two proliferative stimuli on growth factor and immediate early gene expression could be accompanied also by a different effect on GJIC, we analyzed the expression of cx26 and cx32 following PH and LN-induced hyperplasia. The results indicate that while PH, as previously shown, caused a decrease in both CXs, LN induced a striking increase in the levels of cx26 mRNA shortly after treatment (Yamasaki et al. 1993). These results suggest that, depending on the nature of the proliferative stimulus, the expression of Cx32 and Cx26 might be regulated by the same mechanism (PH) or differently (LN).

7.2.4 Liver Ploidy

More recently we have become interested in determining whether the differences observed during the very early events in the cell cycle induced by various different proliferative stimuli could also result in differences during the late phases of the cell cycle. Since no evidence of differences could be found during the "S" phase, and mitosis consistently followed DNA synthesis, we have studied the effect of various proliferative stimuli on the ploidy state of the liver following compensatory regeneration induced by PH or after mitogen-induced liver growth. Normal adult rat liver is constituted predominantly by 4c cells (90%). A small percentage of diploid cells (8%–10%) and octoploid cells (1%) is also present. Among the 4c cells, we have 60% mononucleate hepatocytes and 20% binucleate hepatocytes (BN) which are absent immediately after birth and appear at about 3 weeks after birth, preceding the occurrence of mononucleate tetraploid (MT) cells (Nadal and Zajedela 1966; Wheatley 1972).

After PH there is a rapid decrease in the number of BN cells with an almost complete disappearance 3 days after operation (See Table 1). At the same time an increase in the number of MT and mononucleate

Table 1. Effect of PH and LN on hepatic cellular ploidy and nuclearity

Treatment	2c	2 × 2c	4c	4 × 2c	8c	8 × 2c
Control	10 ± 3	16 ± 1	65 ± 3	6 ± 1	3 ± 0.5	0.1 ± 0.1
PH	9 ± 5	1 ± 0.1	79 ± 1	1 ± 0.5	10 ± 5	0
LN	8 ± 4	11 ± 4	44 ± 1	25 ± 5	8 ± 2	3.7 ± 1.2

Cell proliferation was induced by 70% partial hepatectomy (PH) or lead nitrate (LN; 100 µmol/kg, i.v.). Rats were killed 72 h later. Ploidy and nuclearity were measured using a computer-assisted imaging system in hepatocytes isolated by collagenase perfusion. Each value is the mean ± SD of three to seven rats per group.

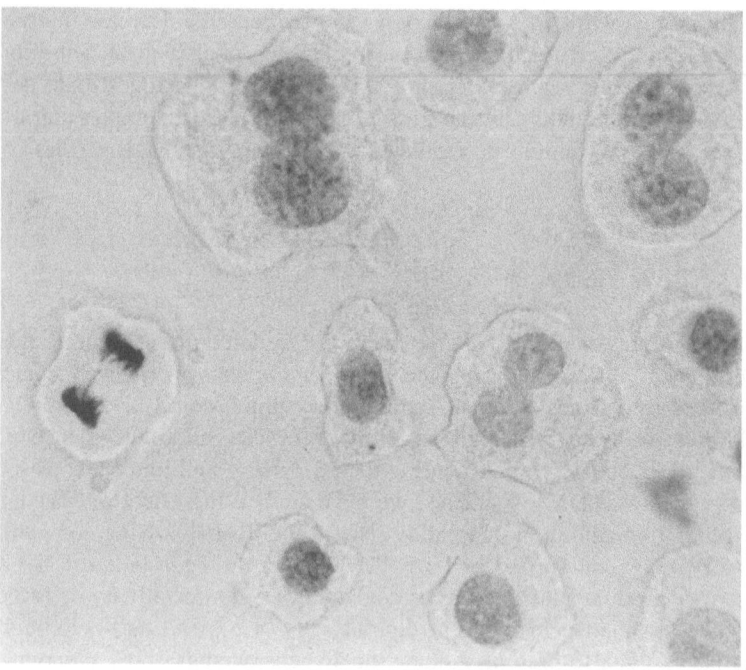

Fig. 2. Binucleated hepatocytes isolated by collagenase perfusion from rats sacrificed 48 h after lead nitrate administration LN (100 µmol/kg). One mitotic phase can also be seen

octaploid (MO) cells is also observed. On the other hand, the ploidy pattern of the hepatocytes after LN is different. The most striking effect is a doubling in the population of BN (Table 1; Fig. 2). Interestingly, the increase is due to 4 × 2c and 8 × 2c hepatocytes, the incidence of 2 × 2c cells being decreased. Thus, while liver regeneration is associated with the disappearance of BN, LN-induced hyperplasia is accompanied by an increase in the percentage of these cells (Melchiorri et al. 1993). A disturbance in cytokinesis seems to be a major phenomenon occurring during LN-induced cell proliferation. It remains to be elucidated whether this is due to the induction by LN of chalone-like factors that specifically inhibit cytokinesis (Nadal 1975) or to alterations in the expression and/or phosphorylating/dephosphorylating state of proteins which play a crucial role in the cytokinetic process (Satterwhite et al. 1992). Alterations in myosin gene are known to cause severe disturbance in the process of cytokinesis and to generate multinucleate cells (Knecht and Loomis 1987; De Lozanne and Spudich 1987).

Irrespective of the mechanism responsible for the changes observed, it is clear that liver growth induced by the mitogen LN induces a ploidy pattern different from that induced by a regenerative stimulus such as PH. However, this phenomenon does not appear to be a common feature during mitogen-induced liver growth. In fact, a single dose of NAF (200 mg/kg) which leads to a shift towards higher ploidy classes, does not modify the percentage of BN cells (Melchiorri et al. 1993). In addition studies by others (Schulte-Hermann et al. 1980; Styles et al. 1990) have shown that liver mitogens such as CPA or methylclofenapate (MCP) induce a disappearance of BN cells, similarly to what seen after PH. Although more studies are needed to determine what is responsible for the different patterns of ploidy seen after different proliferative stimuli, it is clear that a disappearance of BN cells is not necessarily associated to liver cell proliferation induced by acute treatment with direct mitogens.

7.3 Mitogen-Induced Liver Growth and Apoptosis

A rapid regression of the excess liver tissue occurs as soon as the mitogenic stimulus provided by LN is withdrawn. The deletion of excess cells occurs through a particular mode of cell death, namely, apoptosis, characterized by several characteristic biochemical and morphological features (Columbano et al. 1985). The fact that apoptotic cell death does not occur soon after LN treatment but only when DNA content has reached its maximum and stops once the DNA content has regained its original value suggests that under our experimental conditions, apoptosis plays a homeostatic role (i.e., removal of excess cells generated by the mitogenic event) and it is not the consequence of the toxicity of the chemical.

Recently, however, it was suggested that apoptosis may occur as a consequence of a defective cell cycle (Colombel et al. 1992). In particular, the finding that in the prostate, following castration, there is an increased entry into "S" phase not followed by mitotic activity has led to the hypothesis that cells stimulated to enter "S" phase die immediately after their entry because they lack in some critical factor(s). The concept of apoptosis as the result of a defective cell cycle has received support by the finding that the expression of the wild type of p53 mRNA (that has been implicated to mediate apoptosis in other cell types), while it is not induced during normal proliferation of prostate cells it was found to be increased during the apoptotic cell death occurring after the initial proliferation following castration. The concept of apoptosis as the consequence of a defective cell cycle, however, does not appear to be valid in our experimental conditions. In fact, LN-induced DNA synthesis is followed by an increased DNA content and by an increase in mitotic activity, clearly indicating that cell death does not occur during the cell cycle but after its completion. In addition, liver apoptosis occurring during the regression of hyperplasia, unlike that seen in the prostate following castration (Buttyan et al. 1989; Legerm 1987), is not associated with an increased expression of p53 or testosterone-repressed prostate message-2 (TRPM2) mRNA expression (Ledda-Columbano et al. 1994; Helvering et al. 1993). Thus, we believe that under our experimental conditions apoptosis is part of a homeostatic mechanism triggered in order to eliminate the excess tissue. As regeneration following PH or necrosis is a "physiological response" to a pathological condition (cell loss), we may consider the apoptotic process under these conditions as a "physiological process" in response to a pathological condition (cell proliferation occurring in response to xenobiotics).

7.4 Compensatory Regeneration, Mitogen-Induced Liver Growth, and Hepatocarcinogenesis

7.4.1 Initiation

Based on the several differences exhibited by the two proliferative stimuli, we have been interested in the question of whether various proliferative stimuli could affect the carcinogenic process in a different way. It is well known that cell proliferation is associated with the carcinogenic process in a variety of conditions. In the liver, in particular, it appears that carcinogens induce initiation only when their administration is coupled with a proliferative stimulus. Although the mechanism by which cell proliferation plays an important role in initiation is not clear, its involvement in events such as fixation of a miscoding lesion in the newly made DNA may explain its role in this process (Cayama et al. 1978; Ishikawa et al. 1980; Columbano et al. 1981). In most of the studies aimed to determine the role of cell proliferation in the initiation step of chemical hepatocarcinogenesis the proliferative stimulus was achieved by a compensatory type of cell proliferation, such as that occurring after PH, a necrogenic dose of CCl4 or a necrogenic dose of the initiating carcinogen itself. Therefore, it was of interest to us to determine whether mitogen-induced liver growth could exert the same effect on initiation of the carcinogenic process. The results of our studies show that initiation of rat liver carcinogenesis was achieved only when nonnecrogenic doses of carcinogens such as N-methyl-N-nitrosourea, diethylnitrosamine, and benzo(a)pyrene were coupled with compensatory regeneration (PH or CCl4), but not when coupled with cell proliferation induced by direct mitogens (LN, EDB, NAF, CPA) (Columbano et al. 1987; Ledda-Columbano et al. 1989). Similar findings were obtained whether the initiated hepatocytes were monitored as enzyme-altered foci using the resistant hepatocyte model or using the phenobarbital or the orotic acid model. Thus, it appears that only compensatory liver cell proliferation, but not direct hyperplasia induced by mitogens, supports carcinogen-induced initiation.

Table 2. Effect of different proliferative stimuli given to rats fed 2-acetylaminofluorene, orotic acid, or phenobarbital on the development of GGT⁺ foci

Treatment	Proliferative stimulus	Promoting regimen	GGT⁺ foci (number/cm²)
DENA	None	2-AAF	19 ± 4
DENA	PH	2-AAF	62 ± 8
DENA	LN	2-AAF	14 ± 3
DENA	None	OA	1 ± 0.2
DENA	PH	OA	9 ± 3
DENA	LN	OA	2 ± 0.2
DENA	None	PB	3 ± 1
DENA	PH	PB	13 ± 1
DENA	LN	PB	3 ± 1.1

Following initiation with diethylnitrosamine (DENA; 150 mg/kg, i.p.) rats were divided in three groups. The first group was exposed to a diet containing 0.03% 2-aminofluorene (2-AAF). One week later a cell proliferative stimulus was provided by 70% partial hepatectomy (PH) or lead nitrate (LN; 100 μmol/kg, i.v.). Rats were maintained for 1 more week on 2-AAF and then placed on a basal diet and killed 1 week later. Rats of the second and third group were exposed to PH or LN following 2 weeks of feeding with 1% orotic acid or 0.5% PB respectively, and were killed 2 weeks later. Values represent the mean ± SE of six to seven rats per group.

7.4.2 Promotion

A second site at which cell proliferation seems to exert a critical effect is in the promotion of carcinogen-initiated cells. Indeed, an increase in the number of preneoplastic and neoplastic lesions has been demonstrated in the liver of rats subjected to proliferative stimuli such as those elicited by repeated 2/3 PH (Pound and McGuire 1978a). Similarly, multiple treatment with CCl₄, which results in liver damage followed by regeneration, increases the number of liver tumors in mice and rats pretreated with diethylnitrosamine (DENA; Pound and McGuire 1978b; Dragani et al. 1986). In contrast, we have found that while multiple treatment with CCl₄ did promote the growth of DENA-initiated cells to foci and/or nodules, repeated proliferative stimuli induced in rat liver by the mitogens LN and EDB did not exert any promoting effect (Columbano et al. 1990). The fact that the extent of DNA synthesis after the last proliferative stimulus induced by LN was found to be similar to that seen in CCl₄-treated rat liver not only is suggestive of a qualitative

Table 3. Effect of proliferative stimuli of different nature given during 2-AAF feeding on the incidence of GGT[+] foci

Treatment	Proliferative stimulus	GGT[+] foci/cm^2
DENA + 2-AAF	None	8.3 ± 2.6
DENA + 2-AAF	PH	40.2 ± 4.2
DENA + 2-AAF	CCl4	45.4 ± 3.7
DENA + 2-AAF	LN	6.4 ± 3.1
DENA + 2-AAF	EDB	11.1 ± 4.7
DENA + 2-AAF	CPA	11.6 ± 4.2
DENA + 2-AAF	NAF	10.4 ± 5.2

Following initiation with diethylnitrosamine (DENA; 150 mg/kg, i.p.) rats were fed a diet containing 0.03% 2-aminofluorene (2-AAF). After 1 week of 2-AAF feeding, a cell proliferative stimulus was provided by partial hepatectomy (PH), CCl4 (2 mg/kg, per os), lead nitrate (LN; 100 μmol/kg, i.v.), ethylene dibromide (EDB; 100 mg/kg, per os), cyproterone acetate (CPA; 60 mg/kg, per os) or nafenopin (NAF; 200 mg/kg, per os). Rats were maintained for 1 more week on 2-AAF and then placed on a basal diet and killed 1 week later.

rather than quantitative difference between these two proliferative stimuli, but also clearly indicated that DNA synthesis per se is not sufficient to act as a promoting stimulus.

It is known that compensatory regeneration is either a necessary component (resistant hepatocyte model) or it potentiates the promoting ability of agents such as orotic acid and phenobarbital (Columbano et al. 1982; Tatematsu et al. 1989; Ford and Pereira 1980). Therefore, we have recently investigated whether induction of mitogen-induced liver growth in the presence of a promoting environment could enhance the growth of initiated cells to a focal or nodular stage. The results shown in Table 2 indicate that LN-induced proliferation could not substitute for PH or CCl4 in any of the promoting protocols used (Ledda-Columbano et al. 1992). In addition, the inability to support promotion does not appear to be restricted to LN since similar results could be obtained with at least three other mitogens (See Table 3).

7.5 Conclusions

Cell proliferation has long been associated to a carcinogenic process by chemicals. Based on findings that several nongenotoxic agents which induce hepatic growth also induce liver tumor formation, it has been proposed that this increased cell proliferation per se plays a major role in tumor formation by increasing the rate of spontaneous mutations (Ames and Gold 1990; Cohen and Ellwein 1990). However, in spite of the fact that many chemicals that induce liver growth also enhance or accelerate liver tumor formation, a clear-cut correlation between cell proliferation in a given organ and tumor formation has not yet been established (Ledda-Columbano et al. 1989; Eldridge et al. 1992; Melnick 1992), and controversy still exists as to whether peroxisome proliferator-induced liver carcinogenicity is due to a persistent increase in hepatocyte proliferation (Marsman et al. 1988), to oxidative stress (Rao and Reddy 1987), or to other causes. In addition, it is also known that most of the nongenotoxic mitogenic compounds, with very few exceptions (Marsman et al. 1988; Masuhara et al. 1992), induce only a very transient increase in cell proliferation (Yeldandi et al. 1989; Eacho et al. 1991). As described in the present work, the pattern appears to be further complicated by the finding that the nature of the proliferative stimulus (compensatory regeneration versus direct hyperplasia induced by mitogens) may represent an important factor in the carcinogenic process in the liver. At the present time we do not know why the two proliferative stimuli have a different effect on the carcinogenic process. Two possible explanations might be proposed: (1) initiated cells can efficiently respond to regenerative stimuli, but not to stimuli induced by primary mitogens or (2) initiated cells respond to mitogens as efficiently as normal hepatocytes, but are eliminated (together with normal excess cells) by apoptosis during the regression phase of liver hyperplasia.

More work is needed to elucidate the role of cell proliferation in chemically induced hyperplasia. Studies aimed to investigate on the mechanisms underlying the differences between compensatory regeneration and mitogen-induced liver growth may contribute to a better understanding of the role of cell proliferation in the carcinogenic process by chemicals.

Acknowledgments. This work was supported by C.N.R. (Progetto ACRO), A.I.R.C., and MURST (60%) Italy.

References

Ames BN, Gold LS (1990) Too many rodent carcinogens: mitogenesis increases mutagenesis. Science 249:970–971

Asami O, Ihara I, Shimidzu N, Tomita Y, Ichihara A, Nakamura T (1991) Purification and characterization of hepatocytes growth factor from injured liver of carbon tetrachloride-treated rats. J Biochem 109:8–13

Beyer HS, Stanley M, Theologides A (1990) Tumor necrosis factor-alfa increases hepatic DNA and RNA and hepatocyte mitosis. Biochem Int 22:405–410

Beyer EC, Paul DL, Goodenough DA (1990) Connexin family of gap junction proteins. J Membr Biol 116:187–194

Buttyan R, Olsson CA, Pintar J, Chang C, Bandyk M, Ng P-Y, Sawczuk IS (1989) Induction of the TRPM-2 gene in cells undergoing programmed cell death. Mol Cell Biol 9:3473–3481

Cayama E, Tsuda H, Sarma DSR, Farber E (1978) Initiation of chemical carcinogenesis requires cell proliferation. Nature 275:60–61

Cohen S, Ellwein LB (1990) Cell proliferation in carcinogenesis. Science 249:972–975

Colombel M, Olsson CA, Ng P-Y, Buttyan R (1992) Hormone regulated apoptosis results from reentry of differentiated prostate cell onto a defective cell cycle. Cancer Res 52:4313–4319

Columbano A, Rajalakshmi S, Sarma DSR (1981) Requirement of cell proliferation for the initiation of liver carcinogenesis as assayed by three different procedures. Cancer Res 41: 2079–2083

Columbano A, Ledda GM, Rao PM, Rajalakshmi S, Sarma DSR (1982) Dietary orotic acid, a new selective growth stimulus for carcinogen altered hepatocytes in rat. Cancer Lett 16:191–196

Columbano A, Ledda GM, Sirigu T, Perra T, Pani P (1983) Liver cell proliferation induced by a single dose of lead nitrate. Am J Pathol 110:83–88

Columbano A, Ledda-Columbano, GM, Coni P, Faa G, Liguori C, Santacruz G, Pani P (1985) Occurrence of cell death (apoptosis) during the involution of liver hyperplasia. Lab Invest 52:670–675

Columbano A, Ledda-Columbano GM, Lee G, Rajalakshmi S, Sarma DSR (1987) Inability of mitogen-induced liver hyperplasia to support the induction of enzyme-altered islands induced by liver carcinogens. Cancer Res 47:5557–5559

Columbano A, Ledda-Columbano GM, Curto M, Ennas GM, Coni P, Sarma DSR, Pani P (1990) Cell proliferation and promotion of rat liver carcinogenesis: different effect of hepatic regeneration and mitogen-induced hyperplasia on the development of enzyme altered foci. Carcinogenesis 11:771–776

Coni P, Pichiri-Coni G, Ledda-Columbano GM, Rao PM, Rajalakshmi S, Sarma DSR, Columbano A (1990) Liver hyperplasia is not necessarily associated with increased expression of c-fos and c-myc mRNA. Carcinogenesis 11:835–839

Coni P, Pichiri-Coni G, Ledda-Columbano GM, Semple E, Rajalakshmi S, Rao PM, Sarma DSR, Columbano A (1992) Stimulation of DNA synthesis by rat plasma following in vivo treatment with three liver mitogens. Cancer Lett 61:233–238

Coni P, Simbula G, Carcereri De Prati A, Menegazzi M, Sarma DSR, Ledda-Columbano GM, Columbano A (1993) Differences in the steady state levels of c-fos, c-jun and c-myc mRNA during mitogen induced liver growth and compensatory regeneration. Hepatology 17:1109–1116

De Lozanne A, Spudich JA (1987) Disruption of the Dictyostelium myosin heavy chain by homologous recombination. Science 23:1086–1091

Dermietzel R, Yancey SB, Traub O, Willecke K, Revel JR (1987) Major loss of the 28-kD protein of gap junction in proliferating hepatocytes. J Cell Biol 105:1925–1934

Dragani TA, Manenti G, Della Porta G (1986) Enhancing effect of carbon tetrachloride in mouse hepatocarcinogenesis. Cancer Lett 31:171–179

Eacho PI, Lanier TL, Brodhecker CA (1991) Hepatocellular DNA synthesis in rats given peroxisome proliferating agents: comparison of WY-14,643 to clofibric acid, nafenopin and LY171883. Carcinogenesis 12:1557–1561

Eldridge SR, Goldsworthy TL, Popp JA, Butterworth BE (1992) Mitogenic stimulation of hepatocellular proliferation in rodents following 1,4-dichlorobenzene administration. Carcinogenesis 13:409–415

Fausto N, Mead JE (1989) Regulation of liver growth: protooncogenes and transforming growth factors. Lab Invest 60:4–13

Feingold KK, Barker MR, Jones AL, Greenfeld (1991) Localization of tumor necrosis factor stimulated DNA synthesis in the liver. Hepatology 13:773–779

Ford JC, Pereira MA (1980) Short term in vivo initiation/promotion bioassay for hepatocarcinogens. J Environ Pathol Toxicol 4:39–46

Goyette M, Petropulos CJ, Shank PR, Fausto N (1983) Expression of cellular oncogene during liver regeneration. Science 219:510–512

Helvering LM, Richardson KK, Horn DM, Wightman KA, Hall RL, Smith WC, Engelhardt JA, Richardson FC (1993) Expression of TRPM-2 during involution and regeneration of the rat liver. Proc Am Assoc Cancer Res 34:673

Hertzberg E, Johnson R (1988) Gap junctions. Liss, New York (Modern cell biology, vol 7)

Ishikawa T, Takayama S, Kitagawa T (1980) Correlation between time of partial hepatectomy after a single treatment with diethylnitrosamine and induction of adenosine triphosphatase-deficient islands in rat liver. Cancer Res 40:4261–4264

Knecht DA, Loomis WF (1987) Antisense RNA inactivation of myosin heavy chain expression in Dictyostelium discoideum. Science 230:1081–1086

Ledda-Columbano GM, Columbano A, Curto M, Ennas MG, Coni P, Sarma DSR, Pani P (1989) Further evidence that mitogen-induced cell proliferation does not support the formation of enzyme-altered islands in rat liver by carcinogens. Carcinogenesis 10:847–850

Ledda-Columbano GM, Coni P, Curto M, Giacomini L, Faa G, Sarma DSR, Columbano A (1992) Mitogen-induced liver hyperplasia does not substitute for compensatory regeneration during promotion of chemical hepatocarcinogenesis. Carcinogenesis 13:379–383

Ledda-Columbano GM, Coni P, Columbano A (1994) Cell proliferation and cell death in rat liver carcinogenesis by chemicals. Arch Toxicol [Suppl] 18:271–280

Legerm JG, Montpetit ML, Tenniswood MP (1987) Characterization and cloning of androgen-repressed mRNAs from rat ventral prostate. Biochem Biophys Res Commun 147:196–203

Makino R, Hayashi K, Sugimura T (1984) C-myc transcript is induced in rat liver at a very early stage of regeneration or by cycloheximide treatment. Nature 310:697–698

Marsman DS, Cattley RC, Conway JG, Popp JA (1988) Relationship of hepatic peroxisome proliferation and replicative DNA synthesis to the hepatocarcinogenicity of the peroxisome proliferators di(2-ethylhexyl)phtalate and (4-chloro-6(2,3 xylidino)-2-pyrimidinylthio)acetic acid (Wy-14,643) in rats. Cancer Res 48:6739–6744

Masuhara M, Katyal SK, Nakamura T, Shinozuka H (1992) Differential expression of hepatocyte growth factor, transforming growth factor-alfa and transforming growth factor-beta messenger RNAs in two experimental models of liver cell proliferation. Hepatology 16:1241–1249

Matsumoto K, Nakamura T (1991) Hepatocyte growth factor-molecular structure and implications for a central role in liver regeneration. J Gastroenterol Hepatol 6:509–519

Matsumoto K, Tajima H, Hamanoue M, Kohono S, Kinoshita T, Nakamura T (1992) Identification and characterization of "Injurin", an inducer of expression of the gene for hepatocyte growth factor. Proc Natl Acad Sci USA 89:3800–3804

Melchiorri C, Chieco P, Zedda AI, Coni P, Ledda-Columbano GM, Columbano A (1993) Ploidy and nuclearity of rat hepatocytes after compensatory regeneration or mitogen-induced liver growth. Carcinogenesis 14:1825–1830

Melnick RD (1992) Does chemically induced hepatocyte proliferation predict liver carcinogenesis? FASEB J 6:2698–2706

Michalopoulos GK (1990) Liver regeneration: molecular mechanism of growth control. FASEB J 4:240–249

Morello D, Lavenu A, Bobinet A (1990) Differential regulation and expression of c-jun, c-fos and c-myc protooncogenes during mouse liver regeneration and after inhibition of protein synthesis. Oncogene 5:1511–1519

Nadal C (1975) Inhibition of rat hepatocyte multiplication by serum factors. Virchows Arch [B] Cell Pathol 18:273–280

Nadal C, Zajedela F (1966) Polyploidie somatique dans le foie de rat. I. Le role des cellules binuclees dans le genese des cellules polyploides. Exp Cell Res 42:99–116

Neveu MI, Hully JR, Paul DL, Pitot HC (1990) Reversible alteration in the expression of the gap junctional protein connexin 32 during tumor promotion in rat liver and its role during cell proliferation. Cancer Commun 2:21–31

Pound AW, McGuire LJ (1978a) Repeated partial hepatectomies as a promoting stimulus for carcinogenic response of liver to nitrosamines in rats. Br J Cancer 37:585–594

Pound AW, McGuire LJ (1978b) Influence of repeated liver regeneration on hepatic carcinogenesis by diethylnitrosamine in mice. Br J Cancer 37:585–594

Rao MS, Reddy JK (1987) Peroxisome proliferation and hepatocarcinogenesis. Carcinogenesis 8:631–636

Satterwhite LL, Lohka MJ, Wilson KL, Schertson TY, Cisek LJ, Cordern JL, Pollard TD (1992) Phosphorylation of myosin-II regulatory light chain by cyclin-p34^{cdc2}. A mechanism for the timing of cytokinesis. J Cell Biol 118:595–605

Schulte-Hermann R, Hoffmann V, Landgraf H (1980) Adaptive responses of rat liver to the gestagen and anti-androgen cyproterone acetate and other inducers. III: Cytological changes. Chem-Biol Interact 31:301–311

Shinozuka H, Kubo Y, Katyal SL, Coni P, Ledda-Columbano GM, Columbano A (1993) Roles of hepatocyte growth factor and tumor necrosis factor-alfa on liver cell proliferation in rats induced by lead nitrate. FASEB J 7:A429

Styles JA, Bybee A, Pritchard NR, Kelly MD (1990) Studies on the hyperplastic responsiveness of binucleated rat hepatocytes. Carcinogenesis 11:1149–1152

Tatematsu M, Nakanishi K, Murasaki G, Miyata Y, Hirose M, Ito N (1989) Enhancing effect of inducers of liver microsomal enzymes on induction of hyperplastic liver nodules by N2-fluorenylacetamide. J Natl Cancer Inst 63:1411–1416

Traub O, Look J, Dermietzel R, Brummer F, Hulser D, Willecke K (1989) Comparative characterization of the 21-kD and 26-kD gap junction proteins in murine liver and cultured hepatocytes. J Cell Biol 108:1039–1051

Wheatley DN (1972) Binucleation in mammalian liver. Studies on the control of cytokinesis in vivo. Exp Cell Res 74:445–465

Wyllie AH, Kerr JFR, Currie AR (1980) Cell death: the significance of apoptosis. Int Rev Cytol 68:251–306

Yamasaki H, Krutkovskikh V, Mesnil M, Columbano A, Tsuda H, Ito N (1993) Gap junctional intercellular communication and cell proliferation during rat liver carcinogenesis. Environ Health Perspect 109:191–198

Yeldandi AV, Milano M, Subbarao V, Reddy JK, Rao MS (1989) Evaluation of liver cell proliferation during ciprofibrate-induced hepatocarcinogenesis. Cancer Lett 47:21–27

8 The Role of Genotoxic and Nongenotoxic Agents in Multistage Carcinogenesis of Mouse Skin

A. Balmain, C. J. Kemp, P. A. Burns, R. Bremner, S. Bryson,
M. Clarke, S. Williamson and K. Brown

8.1 Introduction

The successful completion of each stage in the multistep transformation
of a normal cell to malignancy involves a number of intrinsically unlikely events. A limited number of target genes, either proto-oncogenes
or tumour suppressor genes, require to be mutated in the appropriate
manner to confer a selective growth advantage on the target cell. Such
mutational events can either take place spontaneously or they may be
induced by exposure to chemical or physical mutagens. In the latter
case, the cell must escape the direct toxic effects of the carcinogenic
agent and in some way circumvent its own attempts to enter a suicide
pathway, which is a frequent primary response to DNA-damaging
agents. All of this must, of course, take place within the correct target
cell, i.e., either a stem cell with a naturally extended lifespan, or a more
committed cell which can acquire an extended lifespan upon exposure
to a carcinogenic stimulus.

Genetic Changes in Skin Carcinogenesis

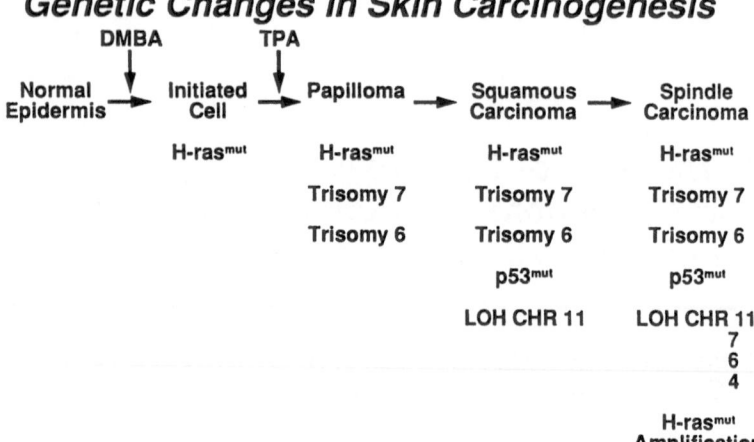

Fig. 1. Scheme showing the main steps involved in mouse skin carcinogenesis. The cumulative genetic changes which have been detected in tumours are shown below the appropriate stage of tumour development. See text for further details. *DMBA*, dimethylbenzanthracene; *TPA*, 12-0-tetradecanoyl phorbol acetate

For several years we have been attempting to identify the target genes which are critically altered at the stages of initiation, promotion or progression of carcinogenesis, using mouse skin as a model system (Balmain et al. 1992). These genetic approaches were designed to complement the extensive biological experiments which have been carried out over the past 50 years on mouse skin tumour development, which have led to many of the currently accepted concepts of multistage carcinogenesis (Yuspa and Poirier 1988). The approaches involved identification of the genes which are mutated at particular stages of tumour development, followed by, under ideal circumstances, a demonstration that the genetic events detected are indeed causal events in the carcinogenesis process. If such rate-limiting steps can be identified, this would offer great encouragement to proceed to the next stage, which is that of intervention in order to prevent or reverse one or more steps of tumour development.

The main features of multistep skin carcinogenesis are depicted in Fig. 1. The initiation step involves a single treatment of mouse skin with a low dose of a genotoxic agent such as dimethylbenzanthracene (DMBA), or methyl-nitro-nitrosoguanidine (MNNG). Subsequent twice-weekly treatment with a tumour promoter such as 12-0-tetradeca-noyl phorbol acetate (TPA) results in the development of benign papillomas, some of which are capable of undergoing an additional progression event to become squamous cell carcinomas. This progression event can itself be accelerated by treatment of benign tumours with mutagenic agents such as MNNG (Hennings et al. 1983). A more recently identified additional progression step involves the conversion of squamous cells to undifferentiated spindle cell carcinomas which are highly invasive and have lost many of the typical differentiation properties of normal epithelial cells (Klein-Szanto 1989; Buchmann et al. 1991; Stoler et al. 1993).

8.2 Allelotype Analysis of Mouse Skin Tumours

Many of the genes which are critically altered in human tumours have been identified using a combination of transfection assays to detect activated oncogenes and allelotype analysis to pinpoint the possible genomic locations of putative tumour suppressor genes (Fearon and Vogelstein 1990). Both of these approaches have been applied to mouse skin tumours. Transfection assays carried out several years ago showed that the majority of both papillomas and carcinomas induced by DMBA and TPA have the same A $\rightarrow$ T transversion mutation at codon 61 of the H-*ras* gene (Quintanilla et al. 1986; Bizub et al. 1986). Strong evidence that the H-*ras* gene constitutes a critical target for initiating action of various mutagens came from studies on mouse skin and on other animal model systems of carcinogenesis (Balmain and Brown 1988; Barbacid 1987; Guerrero and Pellicer 1987; Sukumar 1990). It was shown in several laboratories that altering the nature of the carcinogen used to complete the initiation event led to predictable mutations in the H-*ras* gene in tumours which arose several months after initiation. The identification of *ras* as a critical initiating gene was confirmed in experiments in which the gene itself was used for initiation, either by direct application of a retrovirus to mouse skin (Brown et al. 1986; Roop et al. 1986) or by the introduction of a mutant H-*ras* gene into the germ line of transgenic mice (Bailleul et al. 1990; Greenhalgh et al. 1993). In both

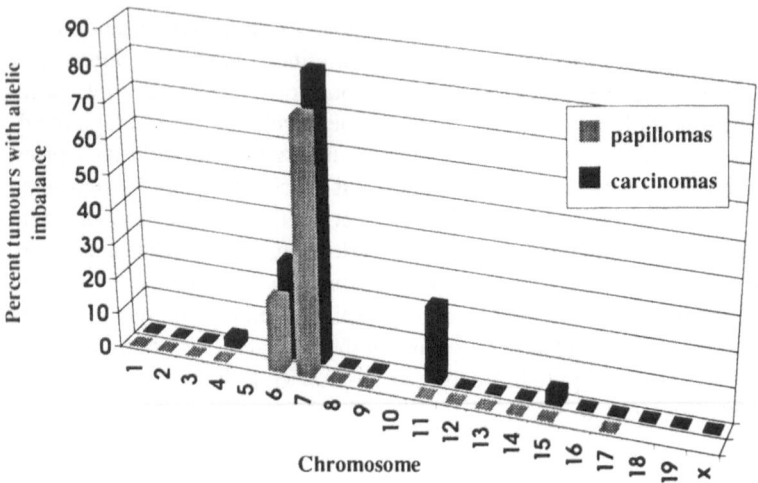

Fig. 2. An allelotype of mouse skin tumours produced using microsatellite markers. See text and Kemp et al. (1993c) for details

cases it could be demonstrated that expression of a mutant *ras* allele within epidermal cells can lead to the formation of benign tumours.

We have now carried out a comprehensive search for additional genetic events which can take place during the promotion and progression stages of tumour development using a panel of microsatellites distributed throughout the mouse genome (Love et al. 1990; Dietrich et al. 1992). This approach was based on the prior development of a hybrid mouse system which would permit studies of loss of heterozygosity in tumours (Bremner and Balmain 1990). Allelic imbalance or loss of heterozygosity could be detected in tumours derived from F1 hybrid mice using markers which demonstrate polymorphism between the parental strains used to generate the hybrids. We have used microsatellites for this analysis because of their high polymorphism rate between different inbred mouse strains and the fact that they are amenable to analysis using the polymerase chain reaction. Over 400 polymorphic microsatellite sequences have been mapped in the mouse genome, approximately 100 of which were used in the present study (Kemp et al. 1993c).

The results obtained using this approach are shown in Fig. 2. At least one marker on 17 of the 19 autosomes was investigated for allelic imbalance in both benign and malignant tumours induced in a variety different F1 hybrid mice. The majority of chromosomes analysed showed no evidence of allelic imbalance. However, it was clear that both chromosomes 6 and 7 showed a high frequency of alterations in benign and malignant tumours. Chromosome 11 and, to a lesser extent, chromosome 4 also showed allelic imbalance, but in this case only in malignant carcinomas.

8.3 Trisomy of Chromosomes 7 and 6 in Papillomas

Previous cytogenetic studies had shown that a trisomy of chromosome 7 can be detected in the majority of chemically induced mouse skin papillomas (Aldaz et al. 1989). The use of a series of markers located at different positions on chromosome 7 confirmed that allelic imbalance could be seen at all of the loci examined, and in each case the appropriate paternal or maternal allele at each locus was over-represented in the tumour (Kemp et al. 1993c). This change was seen in some of the earliest papillomas examined, which were analysed after only 10 weeks post-initiation, indicating that trisomy of chromosome 7 is a very early event in the generation of these tumours.

The mouse H-*ras* gene, which had previously been shown to be mutated at the initiation stage of carcinogenesis, is located at the distal end of mouse chromosome 7 (Kemp et al. 1993a). This suggested that the trisomy might serve to duplicate the mutant *ras* allele, conferring a selective advantage upon the initiated cell. This possibility was indeed confirmed by demonstrating that the chromosome which was duplicated in these tumours was invariably that which carried the mutant H-*ras* gene (Kemp et al. 1993c). This very high frequency of allelic imbalance induced at the early stages of carcinogenesis suggests that there may be a threshold level required for mutant *ras* to confer a selective advantage upon the initiated cell. The single mutant allele which is produced at the time of initiation is obviously insufficient in order to complete this selection process. It has been suggested by others that a critical feature of a tumour promoter such as TPA is its capacity to act as a so-called converting agent for the initiated cell (Furstenberger et al. 1989; Slaga et al. 1980) probably by the induction of genetic instability. It has, indeed, been demonstrated that agents which demonstrably have clastogenic activity, such as methyl methane sulphonate (MMS) can substitute for TPA in this conversion step (Furstenberger et al. 1989), and, moreover, that TPA itself can induce a variety of chromosomal aberrations in target cells, both of fibroblastic (Kinsella and Radman 1978) and epithelial (Petrusevska et al. 1988) origin. Hence, although it is frequently assumed by many workers that TPA is a nongenotoxic agent because of its lack of activity in classical mutagenesis assays, such as the Ames test, it can be seen from these studies that the indirect genotoxic activity of an agent such as TPA, possibly acting via active oxygen species (Emerit and Cerutti 1982), plays a critical role in the early selection of the initiated cells.

There has also been previous cytogenetic evidence that chromosome 6 is trisomic in many mouse skin tumours (Aldaz et al. 1989). From our molecular studies there is clear evidence for trisomy of chromosome 6 in a proportion of papillomas and carcinomas, and for complete loss of heterozygosity on this chromosome in certain tumours. It is not yet known which gene is responsible for this genetic change, but a number of good candidates have emerged. The gene encoding the receptor for hepatocyte growth factor c-*met*, the K-*ras* proto-oncogene, TGF-α and c-*raf*–1 are all localised on chromosome 6 (Elliot and Moore 1992). TGF-α is known to be induced by mutant *ras* alleles or by TPA in epithelial cells, and treatment of mouse epidermal cells with this growth factor can reproduce some of the features of tumour growth (Pittelkow et al. 1989; Finzi et al. 1988; Cheng et al. 1993). Another intriguing possibility is that the crucial target gene on chromosome 6 is the c-*raf*–1 proto-oncogene. This gene has recently been shown to act downstream of *ras* in the signalling pathway which leads from membrane bound tyrosine kinase receptors to the nucleus (for review, see Medema and Bos 1993). One possibility is that duplication of the mutant H-*ras* allele by trisomy of chromosome 7 may require a simultaneous increase in c-*raf*–1 expression in order to maximise the signal output. In vitro experiments have indeed recently demonstrated cooperation between *raf* and *ras* in the transformation of NIH 3T3 cells (Cuadrado et al. 1993).

8.4 Genetic Alterations at the Benign to Malignant Transition

The most frequent genetic change which was specific to carcinomas was loss of heterozygosity on mouse chromosome 11. The presence of the p53 tumour suppressor gene on this chromosome suggested that this gene was the most likely target for mutation at the benign to malignant transition. Preliminary studies (Burns et al. 1991; Ruggeri et al. 1991) demonstrated that the p53 gene can be mutated in a number of chemically induced skin carcinomas. We adopted two different approaches to test the hypothesis that p53 is crucial for malignant progression. The first approach involved examination of the p53 mutation spectrum in tumours induced by different carcinogenesis protocols. It has previously been demonstrated by others that tumour progression is a separate genetic event which can be induced by treatment of benign tumours with mutagenic agents (Hennings et al. 1983). We therefore induced mouse skin carcinomas by complete carcinogenesis protocols using multiple treatments with either DMBA or MNNG. If p53 is a crucial target at this stage of tumour progression, one would expect that the mutation spectrum induced in the p53 gene would differ according to the type of agent used to induce progression. Figure 3 summarises the results of this study and indicates that this hypothesis was indeed verified. Sequencing of p53 alleles from tumours induced by complete carcinogenesis protocols demonstrated two features. First of all, the incidence of loss of heterozygosity was considerably lower than in carcinomas which had been derived by the classical initiation–promotion protocol. Secondly, the mutations detected in the p53 gene varied according to the type of agent used, with transversion mutations, predominantly $A \rightarrow T$ or $G \rightarrow T$ being observed, in the DMBA-induced tumours and $G \rightarrow A$ transition mutations in those produced by multiple MNNG exposure. These particular mutation types i.e. transversion or $G \rightarrow A$ transition mutations respectively are those expected from classical mutagenesis studies using these mutagens. In fact the *ras* gene itself exhibits a very similar spectrum of mutations when initiation is carried out with the same chemical carcinogens (Fig. 3).

The second approach employed was to investigate the progression frequency of mouse skin tumours induced in mice which completely lack a functional p53 gene. Such mice have been produced by homolo-

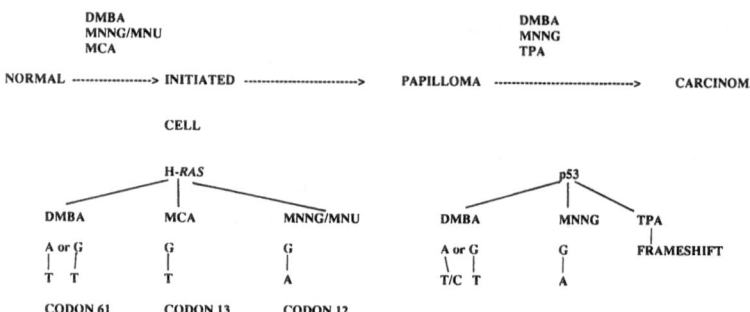

Fig. 3. *Ras* and p53 genes as targets for mutagenesis by chemical carcinogens. The types of mutations seen in the *ras* or p53 genes in tumours produced after treatment with initiating or progressing agents, respectively, are shown at the *bottom*

gous recombination in embryonic stem cells (Donehower et al. 1992). Somewhat unexpectedly, they show normal development but have a high spontaneous tumour incidence between 3 and 6 months of age, leading to the early death of the animals. The majority of tumours which arise are lymphomas and soft tissue sarcomas. Very few spontaneously arising carcinomas have been observed in such animals, possibly due to their early death for other reasons. We therefore carried out a series of initiation–promotion experiments using p53 null mice. The results (Kemp et al. 1993b) demonstrated conclusively that p53 plays a causal role in the progression from benign to malignant tumours. The results obtained using the null animals demonstrated a number of interesting features. First of all, while no difference was detected in the incidence, size, or rate of appearance of benign tumours between normal animals and those heterozygous for the inactivated p53 allele, animals homozygous for the nonfunctional alleles showed a reduced incidence of benign tumours. This unexpected finding may be due to the putative role of the p53 tumour suppressor gene in the control of genetic stability (Lane 1992). It has been postulated that in normal mice the action of genotoxic agents upon a cell induces growth arrest as a prelude to DNA repair. p53 is implicated in the growth arrest process, and its absence in null mice may render some of the exposed cells particularly sensitive to the toxic effects of the chemical carcinogen used. Since successful initiation of carcinogenesis requires misrepair of lesions in DNA, many cells which in wild type animals would become initiated may in null mice make a catastrophic attempt at DNA replication, leading to cell death. We have used similar arguments to explain the relatively low tumour incidence (considering that they have mutant p53 alleles in every cell of the body) in patients exhibiting Li Fraumeni syndrome (Kemp et al. 1993b).

Although the null mice exhibited a relatively low incidence of benign tumours, these progressed to malignancy at extremely high frequency. Whereas under normal circumstances the progression rate from papilloma to carcinoma is of the order of 5%–10% (Hennings et al. 1983), papillomas in null mice progressed at a frequency approaching 50%. This may, in fact, be an underestimate of the true value, since many animals died before the tumours had sufficient time to undergo malignant progression. Histological examination of both benign and malignant tumours induced in these experiments demonstrated that the whole process of malignant progression is vastly accelerated in p53 null mice. While some of the benign papillomas showed all of the classic features seen in papillomas from wild type animals, some of them exhibited areas of increased cellularity and decreased differentiation more typical of carcinoma in situ. These observations have subsequently been confirmed using a number of antibodies to epidermal differentiation markers, including keratins, cell adhesion molecules and transforming growth factor beta (in collaboration with the groups of R.J. Akhurst, Glasgow, A. Cano and M. Quintanilla, Madrid). Among the malignant tumours, several were detected which had a highly invasive spindle cell phenotype and the incidence of lymph node metastases was increased in comparison to tumours induced in wild type mice.

Both of these approaches lead us to the same conclusion, namely that mutational alterations leading to inactivation of the p53 gene are a causal and probably rate-limiting step in the progression of benign to malignant tumours. The accumulated studies on genetic alterations in rodent tumours have reinforced the notion that mechanisms of carcinogenesis in humans are largely similar to those seen in experimental animals. However, whereas in rodents the aetiology of tumour development can be controlled, and the types of agent which can induce particular stages of carcinogenesis or specific genetic alterations are known, in humans these correlations can only be indirectly inferred from the known mechanisms of action of carcinogenic agents in model systems.

Studies on mutagenesis by polycyclic hydrocarbons or alkylating agents, either in bacterial systems or in mammalian cells in vitro, have shown that these mutagens induce predominantly transversions or transition mutations, respectively. A corresponding mutation spectrum is seen in the H-*ras* gene as a function of the type of initiating agent used to induce tumours of the skin, mammary gland or other organs (Balmain and Brown 1988; Barbacid 1987). It has also been noted that Kirsten *ras* mutations in lung cancers from patients who smoke are also predominantly of the G → T transversion type, an observation which has been attributed to the concentration of polycyclic hydrocarbons in cigarette smoke. A different spectrum of mutations is seen in the *ras* genes in tumours of the gastrointestinal tract (Capella et al. 1991).

The p53 gene would also appear to be a direct target for the mutagenic action of environmental carcinogens (Harris 1991). This has been most clearly demonstrated for tumours of the skin, where Brash and colleagues have shown that basal and squamous cell carcinomas, known from epidemiological studies to be due to persistent exposure to ultraviolet light, have mutations primarily at dipyrimidine sites (Brash et al. 1991). These types of double mutations (CC → TT transitions) were thought to be highly specific to ultraviolet light exposure, suggesting that p53 is indeed a direct target for mutagenesis by this agent. However, some doubt has been cast on this hypothesis by the observation that under certain circumstances, exposure of DNA to active oxygen species can also induce mutations of this kind (Reid and Loeb 1993). In order to obtain definitive evidence that p53 can indeed be a target for mutation at the time of progression to malignancy, animal models have therefore again been invaluable. The studies described above, in addition to those already published from this and other laboratories (Burns et al. 1991; Ohgaki et al. 1992; Kress et al. 1992), provide definitive proof that the mutation spectrum of the p53 gene is indeed a function of the chemical or physical agent used to induce carcinogenesis.

The role of promoting agents in tumour induction is by comparison much less clear. Promoters are known to induce cell proliferation, and the role of proliferation in carcinogenesis has been the subject of considerable debate (Ames and Gold 1990; Weinstein 1992; Tennant et al. 1991). The so-called nongenotoxic carcinogens are frequently thought to fall into the promoter class. However, the distinction between genotoxic and nongenotoxic agents is becoming increasingly blurred, since tumour promoters such as TPA are known to cause mitotic aneuploidy in yeast (Parry et al. 1981) and can also cause chromosomal alterations in mammalian cells (Kinsella and Radman 1978; Furstenberger et al. 1989). The ability of agents such as TPA to induce gross chromosomal changes of this kind may be a crucial step in the conversion stage of carcinogenesis, as previously suggested by Furstenberger et al. (1989). We have proposed that the induction of nondisjunction events leading to trisomy of chromosomes 7 and possibly 6 is an important event which facilitates the growth of initiated cells. Thus, although this agent is negative in the vast majority of assays for genotoxicity, it clearly plays an important role in carcinogenesis as a consequence of genotoxic activity which may be of a more indirect type. It should also be noted in this context that the stimulation of cell proliferation by treatment with naturally occurring growth factors such as epidermal growth factor (EGF) or by transfection of oncogenes (Kelsey et al. 1987; Stenman et al. 1987) can apparently also induce a low level of genetic instability.

The role of promotion in the induction of human cancers is still unclear. We suggest that studies on animal model systems may help to identify the types of genetic events associated with the promotion process and consequently identify human tumours in which promotion is likely to have played an important part. Such considerations are not trivial, since much is known about the phenomenology of promotion in skin and about the means by which this part of the carcinogenesis process may be inhibited. Any clear indication of the activity of promoting agents in human cancers may therefore identify tumour types which could be inhibited by mechanisms known to be operative in model systems such as skin. The studies described above on mouse skin have clearly shown that tumours induced by initiation followed by promotion have a much higher incidence of chromosomal nondisjunction events generating loss of heterozygosity than tumours induced by repeated treatment with low doses of a carcinogenic agent (Bremner et al. 1994). Since nondisjunction is very frequently seen in tumours of the human colon, leading to homozygosity of mutant p53 alleles, one could speculate that this may be due in part to proliferation induced by dietary promoting agents. An interesting parallel may be made with human skin tumours known to be induced by chronic exposure to the mutagenic effects of ultraviolet light. These show a much lower incidence of loss of heterozygosity on chromosome 17p and a correspondingly higher incidence of double mutations leading to the independent inactivation of the two p53 alleles (Ziegler et al. 1993). This mechanism therefore has more similarities with the results obtained using complete carcinogenesis protocols on mouse skin by chronic treatment with low doses of mutagenic agents. The continued investigation of genetic mechanisms of carcinogenesis in animals will hopefully therefore provide additional information which will be of use in elucidating the aetiology of human cancer and, consequently, in the design of new preventative or therapeutic strategies.

Acknowledgements. The Beatson Institute is supported by grants from the Cancer Research Campaign. CJK was funded by a Fellowship from the International Agency for Research on Cancer, and by a grant to A.B. and Tom Wheldon from the United Kingdom Coordinating Committee on Cancer Research. We are grateful to Stephen Bell for help with the mouse colony.

References

Aldaz CM, Trono D, Larcher F, Slaga TJ, Conti CJ (1989) Sequential trisomization of chromosomes 6 and 7 in mouse skin premalignant lesions. Mol Carcinog 222–26

Ames BN, Gold LS (1990) Too many rodent carcinogens: mitogenesis increases mutagenesis. Science 249:970–971

Bailleul B, Surani MA, White S, Barton SC, Brown K, Blessing M, Jorcano J, Balmain A (1990) Skin hyperkeratosis and papilloma formation in transgenic mice expressing a ras oncogene from a suprabasal keratin promoter. Cell 62:697–708

Balmain A, Brown K (1988) Oncogene activation in chemical carcinogenesis. Adv Cancer Res 51:147–182

Balmain A, Kemp CJ, Burns PA, Stoler AB, Fowlis DJ, Akhurst RJ (1992) Functional loss of tumour suppressor genes in multistage chemical carcinogenesis. In: Harris CC, Hirohashi S, Ito N, Pitot HC, Sugimura T, Terada M, Yokota J (eds) Multistage carcinogenesis. Japanese Scientific Society Press, Tokyo/CRC Press, Boca Raton, pp 97–108

Barbacid M (1987) ras genes. Annu Rev Biochem 56:779–827

Bizub D, Wood AW, Skalka AM (1986) Mutagenesis of the Ha-ras oncogene in mouse skin tumors induced by polycyclic aromatic hydrocarbons. Proc Natl Acad Sci USA 83:6048–6052

Brash DE, Rudolph JA, Simon JA, Lin A, McKenna GJ, Baden HP, Halperin AJ, Ponten J (1991) A role for sunlight in skin cancer: UV-induced p53 mutations in squamous cell carcinoma. Proc Natl Acad Sci USA 88:10124–10128

Bremner R, Balmain A (1990) Genetic changes in skin tumour progression: correlation between presence of a mutant ras gene and loss of heterozygosity on mouse chromosome 7. Cell 61:407–417

Bremner R, Kemp CJ, Balmain A (1994) Different classes of chemical carcinogens induce different genetic changes during progression of mouse skin tumours. Mol Carcinog (in press)

Brown K, Quintanilla M, Ramsden M, Kerr IB, Young S, Balmain A (1986) v-ras genes from Harvey and BALB murine sarcoma viruses can act as initiators of two-stage mouse skin carcinogenesis. Cell 46:447–456

Buchmann A, Ruggeri B, Klein-Szanto AJP, Balmain A (1991) Progression of squamous carcinoma cells to spindle carcinomas of mouse skin is associated with an imbalance of H-ras alleles on chromosome 7. Cancer Res 51:4097–4101

Burns PA, Kemp CJ, Gannon JV, Lane DP, Bremner R, Balmain A (1991) Loss of heterozygosity and mutational alterations of the p53 gene in skin tumors of interspecific hybrid mice. Oncogene 6:2363–2369

Capella G, Cronauer Mitra S, Pienado MA, Perucho M (1991) Frequency and spectrum of mutations at codons 12 and 13 of the c-K-ras gene in human tumors. Environ Health Perspect 93:125–131

Cheng C, Tennenbaum T, Dempsey PJ, Coffey RJ, Yuspa SH, Dlugosz A (1993) Epidermal growth factor receptor ligands regulate keratin 8 expression in keratinocytes, and transforming growth factor alpha mediates the induction of keratin 8 by the v-ras Ha oncogene. Cell Growth Differ 4:317–327

Cuadrado A, Bruder JT, Heideran MA, App H, Rapp UR, Aaronson SA (1993) H-ras and raf-1 cooperate in transformation of NIH3T3 fibroblasts. Oncogene 8:2443–2448

Dietrich W, Hillary K, Lincoln SE, Shin H-S, Friedman J, Dracopoli NC, Lander ES (1992) A genetic map of the mouse suitable for typing intraspecific crosses. Genetics 131:423–447

Donehower LA, Harvey M, Slagle BL, McArthur MJ, Montgomery CA Jr, Butel JS, Bradley A (1992) Mice deficient for p53 are developmentally normal but susceptible to spontaneous tumours. Nature 356:215–221

Elliot RW, Moore KJ (1992) Mouse chromosome 6. Mammal Gen 3:S81-S103

Emerit I, Cerutti PA (1982) Tumour promoter phorbol 12-myristate 13-acetate induces a clastogenic factor in human lymphocytes. Proc Natl Acad Sci USA 79:7509–7513

Fearon ER, Vogelstein B (1990) A genetic model for colorectal tumorigenesis. Cell 61:759–767

Finzi E, Kilkenny A, Strickland JE, Balaschack M, Bringman T, Derynck R, Aaronson S, Yuspa SH (1988) TGFα stimulates growth of skin papillomas by autocrine and paracrine mechanisms but does not cause neoplastic progresssion. Mol Carcinog 1:7–12

Furstenberger G, Schurich B, Kaina B, Petrusevska RT, Fusenig NE, Marks F (1989) Tumor induction in initiated mouse skin by phorbol esters and methyl methanesulfonate: correlation between chromosomal damage and conversion ('stage I of tumor promotion') in vivo. Carcinogenesis 10:749–752

Greenhalgh DA, Rothnagel JA, Quintanilla MI, Orengo CC, Gagne TA, Bundman DS, Longley MA, Roop DR (1993) Induction of epidermal hyperplasia, hyperkeratosis, and papillomas in transgenic mice by a targeted v-ha-ras oncogene. Mol Carcinog 7:99–110

Guerrero I, Pellicer A (1987) Mutational activation of oncogenes in animal model systems of carcinogenesis. Mutat Res 185:293–308

Harris CC (1991) Molecular basis of multistage carcinogenesis. Princess Takamatsu Symp 22:3–19

Hennings H, Shores R, Wenk ML, Spangler EF, Tarone R, Yuspa SH (1983) Malignant conversion of mouse skin tumours is increased by tumour initiators and unaffected by tumour promoters. Nature 304:67–69

Kelsey KT, Nagasawa H, Umans RS, Little JB (1987) Epidermal growth factor induces cytogenetic damage in mammalian cells. Carcinogenesis 8:625–627

Kemp CJ, Bremner R, Balmain A (1993a) A revised map position for the Ha-ras gene on mouse chromosome 7: implications for analysis of genetic alterations in rodent tumors. Mol Carcinog 7:147–150

Kemp CJ, Donehower LA, Bradley A, Balmain A (1993b) Reduction of p53 gene dosage does not increase initiation or promotion but enhances malignant progression of chemically induced skin tumors. Cell 74:813–822

Kemp CJ, Fee F, Balmain A (1993c) Allelotype analysis of mouse skin tumours using polymorphic microsatellites: stepwise genetic alterations on chromosomes 6, 7 and 11. Cancer Res 53:6022–6027

Kinsella AR, Radman M (1978) Tumour promoter induces sister chromatid exchanges: relevance to mechanisms of carcinogenesis. Proc Natl Acad Sci USA 75:6149–6153

Klein-Szanto AJP (1989) Pathology of human and experimental skin tumors. In: Conti CJ, Slaga TJ, Klein-Szanto AJP (eds) Skin tumors. Experimental and clinical aspects. Raven, New York, pp 19–53 (Carcinogenesis, vol 11)

Kress S, Sutter C, Strickland PT, Mukhtar H, Schweizer J, Schwarz M (1992) Carcinogen-specific mutational pattern in the p53 gene in ultraviolet B radiation-induced squamous cell carcinomas of mouse skin. Cancer Res 52:6400–6403

Lane DP (1992) p53, guardian of the genome. Nature 358:15–16

Love JM, Knight AM, McAleer MA, Todd JA (1990) Towards construction of a high resolution map of the mouse genome using PCR-analysed microsatellites. Nucleic Acids Res 18:4123–4130

Medema RH, Bos JL (1993) The role of p21 ras in receptor tyrosine kinase signalling. Crit Rev Oncog 4:615–661

Ohgaki H, Hard GC, Hirota N, Maekawa A, Takahashi M, Kleihues P (1992) Selective mutation of codons 204 and 213 of the p53 gene in rat tumors induced by alkylating N-nitroso compounds. Cancer Res 52:2995–2998

Parry JM, Parry EM, Barrett JC (1981) Tumour promoters induce mitotic aneuploidy in yeast. Nature 294:263–265

Petrusevska RT, Furstenberger G, Marks F, Fusenig NE (1988) Cytogenetic effects caused by phorbol ester tumor promoters in primary mouse keratinocyte cultures: correlation with the convertogenic activity of TPA in multistage skin carcinogenesis. Carcinogenesis 9:1207–1215

Pittelkow MR, Lindquist PB, Abraham RT, Graves DR, Derynck R, Coffey RJ (1989) Induction of transforming growth factor alpha expression in human keratinocytes by phorbol esters. J Biol Chem 264:5164–5171

Quintanilla M, Brown K, Ramsden M, Balmain A (1986) Carcinogen-specific mutation and amplification of Ha-ras during mouse skin carcinogenesis. Nature 322:78–80

Reid TM, Loeb LA (1993) Tandem double CC–TT mutations are produced by reactive oxygen species. Proc Natl Acad Sci USA 90:3904–3907

Roop DR, Lowy DR, Tambourin PE, Strickland J, Harper JR, Balaschak M, Spangler EF, Yuspa SH (1986) An activated Harvey ras oncogene produces benign tumours on mouse epidermal tissue. Nature 323:822–824

Ruggeri B, Caamano J, Goodrow T, DiRado M, Bianchi A, Trono D, Conti CJ, Klein-Szanto AJP (1991) Alterations of the p53 tumor suppressor gene during mouse skin tumor progression. Cancer Res 51:6615–6621

Slaga TJ, Fischer SM, Nelson K, Gleason GL (1980) Studies on the mechanism of skin tumor promotion: evidence for several stages in promotion. Proc Natl Acad Sci USA 77:3659–3663

Stenman G, Delorme EO, Lau CC, Sager R (1987) Transfection with plasmid pSV2ptEJ induces chromosome rearrangements in CHEF cells. Proc Natl Acad Sci USA 84:184–188

Stoler AB, Stenback F, Balmain A (1993) The conversion of mouse skin squamous cell carcinomas to spindle cell carcinomas is a recessive event. J Cell Biol 122:1103–1117

Sukumar S (1990) An experimental approach to cancer. Cancer Cells 2:99–204

Tennant RW, Elwell MR, Spalding JW, Griesemer RA (1991) Evidence that toxic injury is not always associated with induction of chemical carcinogenesis. Mol Carcinog 4:420–440

Weinstein IB (1992) Toxicity, cell proliferation, and carcinogenesis. Mol Carcinog 5:2–3

Yuspa SH, Poirier MC (1988) Chemical carcinogenesis: from animal models to molecular models in one decade. Adv Cancer Res 50:25–70

Ziegler A, Leffell DJ, Kunala S, Sharma HW, Gailani M, Simon JA, Halperins AJ, Baden HP, Shapiro PE, Bale AE, Brash DE (1993) Mutation hotspots due to sunlight in the p53 gene of nonmelanoma skin cancers. Proc Natl Acad Sci USA 90:4216–4220

9 Liver Tumor Promotion and Breast Cancer Chemoprevention: Common Mechanisms

R. L. Jirtle

9.1 Liver Tumor Promotion

Carcinogenesis has been operationally divided into three stages (Fig. 1) (Dragan and Pitot 1992). The first stage is initiation, which involves an irreversible genetic change that occurs in cells either spontaneously or upon exposure to chemical and physical agents. The reversible process of promotion, the second stage of neoplastic development, selectively expands the number of these initiated cells. The final stage of tumor development is progression. It entails the accumulation of additional genetic lesions that result in the activation of oncogenes and/or the inactivation of tumor suppressor genes (Vogelstein and Kinzler 1993). It is during this latter stage of tumor development that hepatocellular carcinomas are observed forming within benign adenomas.

Though tumor-promoting agents increase both the number and the size of tumors, it must be stressed that tumors are not created in a vacuum, but rather developing liver tumors are surrounded by normal

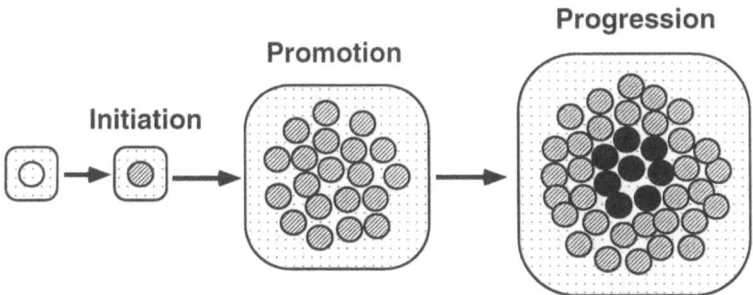

Fig. 1. Multistage carcinogenesis model. Carcinogenesis is divided into three stages: initiation, promotion, and progression. The initiated cells (*hatched circles*) are promoted to form nodules by promoting agents. Ultimately, some of these initiated cells change sufficiently (*filled circles*) to progress to malignant carcinomas

cells. Therefore, to understand the mechanisms by which promoters enhance liver tumor formation, it is important to determine the growth effects these agents have not only on the initiated cells but also on the surrounding normal hepatocytes. For in the liver where homeostatic mechanisms are operating, it is the differential growth effect that liver tumor promoting agents have on the normal versus the initiated cells which provides the selective proliferative pressure that ultimately gives rise to tumor formation.

The first xenobiotic agent shown to promote the formation of liver tumors in rats is the barbituate phenobarbital (PB; Peraino et al. 1971). Though PB has been shown to be a potent promoter of liver tumors in rats and mice, epidemiological studies have not demonstrated a tumor promotional effect in humans (McLean et al. 1992). The reason for this marked species difference in the promotional efficacy of PB is presently unknown, but it demonstrates that a clearer understanding of the mechanisms of tumor promotion is required in order to accurately assess human risk.

When animals are placed on PB the liver increases significantly in size. Part of this increase in liver mass is due to cellular hypertrophy; however, PB also causes a significant degree of hyperplasia, principally in the pericentral region of the liver (Peraino et al. 1975; Yager et al. 1986; Schulte-Hermann et al. 1986; Jirtle et al. 1991). Thus, the short-term effect of PB is to a significantly increase hepatocyte proliferation. This proliferative response, however, maximizes approximately 3 days after starting the exposure of animals to PB, and by 1- to 2-week hepatocyte proliferation has returned to normal (Peraino et al. 1975; Yager et al. 1986). It takes, however, approximately 3 months of continuous exposure to PB to maximally promote liver tumor formation (Preat et al. 1987). Therefore, to more fully understand the mechanisms by which liver tumor promoters enhance tumor formation, it is important to also determine the effect of long-term PB exposure on the proliferative capacity of hepatocytes.

Interestingly, our results (Jirtle et al. 1991) and those of others (Abanobi et al. 1982; Barbason et al. 1983) demonstrate that hepatocyte proliferation in response to mitogenic stimuli is significantly reduced by long-term exposure to PB. We have shown that this reduced ability of PB-treated hepatocytes to proliferate is correlated with a significant reduction in the epidermal growth factor (EGF) receptor number and a loss in the ability of protein kinase C to be translocated to the membrane (Brockenbrough et al. 1991; Jirtle et al. 1991; Jirtle and Meyer 1991). Thus, though PB initially enhances normal hepatocyte proliferation, the long-term PB exposure required for tumor promotion dramatically inhibits the growth potential of normal but not initiated hepatocytes (Jirtle et al. 1991; Jirtle and Meyer 1991).

9.1.1 TGF-β and Liver Tumor Promotion

Transforming Growth Factor-β (TGF-β) has two main biological functions. It stimulates the formation of connective and supporting tissue, and it is also a potent inhibitor of epithelial cell proliferation (for reviews see Massagué 1990; Border and Ruoslahti 1992). Because of TGF-β's potent mitoinhibitory effect on epithelial cells, we investigated the role that TGF-β plays in PB-induced liver tumor promotion.

TGF-β is synthesized as a prepromolecule 390 amino acids in length (Fig. 2). After the signal peptide is removed, two promolecules form a homodimer that is further processed by proteolytic cleavage at a dibasic site 112 amino acids from the C-terminal end; this TGF-β homodimer is held together by a single disulfide bond (Schlunegger and Grütter 1992). The TGF-β molecule, however, remains ionically associated

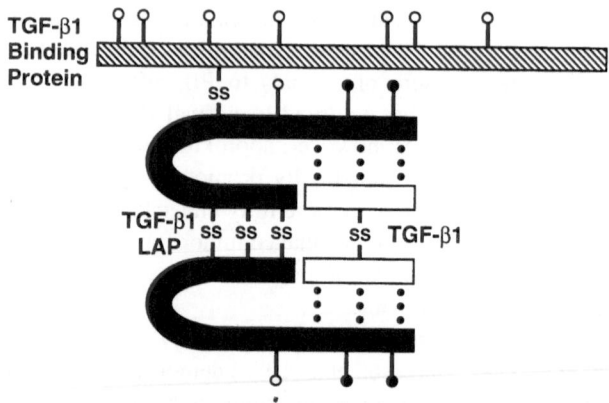

Fig. 2. Model of the large latent transforming growth factor (*TGF-β1*) complex. This model is based upon the results of Kankaki et al. (1990). The *open circles* represent putative N-linked glycosylation sites; those that contain mannose 6-phosphate residues are depicted by *filled circles*. The TGF-β1 latent associated protein *TGF-β1-LAP*, is the N-terminal portion of the TGF-β1 pro molecule. The TGF-β1 binding protein and TGF-β1-LAP are coded for by different genes. *SS*, disulfide bonds

with the N-terminal portion of the promolecule after proteolytic cleavage. Furthermore, this TGF-β latent-associated protein (i.e. TGF-β LAP) has three N-linked glycosylation sites; two of which contain mannose 6-phosphate residues. Though the association of TGF-β with TGF-β LAP is sufficient to confer latency to the TGF-β molecule, a 150-kDa protein (i.e., TGF-β binding protein) that is transcribed from a separate gene is also linked to the TGF-β LAP by disulfide bonds (Kankaki et al. 1990). Though both the large and small latent complex can be formed in cells, it is normally secreted as a large latent complex (Olofsson et al. 1992). Therefore, the biological effectiveness of TGF-β is not only regulated by its synthesis but also by its activation.

Upon release from this latent complex, TGF-β binds to its signaling receptors. Though five receptors have been shown to bind TGF-β , three of these receptors are ubiquitous and have recently been cloned (Ebner et al. 1993; Lin et al. 1992; Lopez-Cassillas et al. 1991; Wang et al. 1991). The type I and type II TGF-β receptors are serine/threonine kinase receptors and are involved in signal transduction. The TGF-β

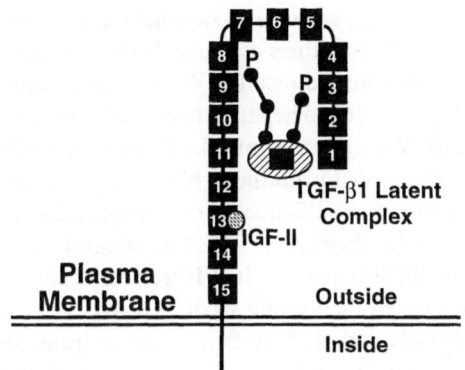

Fig. 3. Model of the cation-independent mannose 6-phosphate/insulin-like growth factor II (M6P/IGF-II) receptor. The M6P/IGF-II receptor contains two different ligand binding sites and binds both mannose 6-phosphate containing glycoproteins and insulin-like growth factor (*IGF-II*). The M6P/IGF-II receptor contains a 2269 residue extracytoplasmic domain, a single 23-residue transmembrane region, and a short 163-residue carboxyl-terminal cytoplasmic domain. The extracytoplasmic domain has a repetitive structure consisting of 15 contiguous repeating segments (*1–15*) of approximately 147 amino acids each. *TGF-β1*, transforming growth factor. *P*, phosphate group attached to a mannose residue

Type I receptor has been shown to be primarily involved in stimulating fibrosis formation, whereas the type II receptor transduces the mitoinhibitory signals of TGF-β (Chen et al. 1993). Interestingly, the type III receptor, also referred to as betaglycan, does not seem to be directly involved in signaling, but rather binds TGF-β with high affinity and presents the TGF-β molecule to the type I and II TGF-β signaling receptors (Lopez-Casillas et al. 1993). Therefore, betaglycan appears to function primarily in enhancing the biological effectiveness of TGF-β.

To investigate the role of TGF-β in liver tumor promotion, we first determined whether TGF-β expression was increased in the liver by PB exposure. To perform these studies, we immunohistochemically stained the liver of PB-treated animals for TGF-β1. We observed that the periportal hepatocytes in PB-treated animals contained significantly elevated levels of TGF-β1 when compared to those in unexposed liver (Jirtle and Meyer 1991). By quantifying TGF-β1 in hepatocytes isolated from PB-treated animals, we have shown that the periportal hepatocytes contain approximately ten times as much TGF-β1 as hepatocytes from control animals (unpublished results). We also determined whether this increase in TGF-β was due to hepatocyte production and/or to increased uptake of TGF-β. We extracted mRNA from isolated PB-treated and control hepatocytes and by northern blot analysis demonstrated that neither the normal nor the PB-treated cells expressed detectable levels of TGF-β1 mRNA (Jirtle et al. 1994). This suggests that the increased concentration of TGF-β present in PB-treated hepatocytes is due to enhanced uptake rather than synthesis of TGF-β1.

As we previously stated, TGF-β is secreted from cells as a latent glycoprotein complex that contains mannose 6-phosphate residues. Furthermore, the TGF-β latent complex has been shown to bind to the mannose 6-phosphate/insulin-like growth factor II (M6P/IGF-II) receptor (Kovacina et al. 1989; Fig. 3). This chimeric receptor has two distinct binding sites, one for IGF-II and another for mannose 6-phosphate containing glycoproteins. The M6P/IGF-II receptor is coded for by a maternally imprinted gene (Barlow et al. 1991; Stöger et al. 1993) that is present on chromosome 17 (region A–C) in the mouse and chromosome 6 (region 6q25–6q27) in humans (Laureys et al. 1988). Thus, the M6P/IGF-II receptor gene contains only a single functional allele. Interestingly, IGF-II is paternally imprinted (Barlow et al. 1991; Stöger et al. 1993), and an evolutionary reason for the opposite imprinting of these two genes has been postulated (Haig and Graham 1991).

The M6P/IGF-II receptor is highly expressed during embryogenesis, but except for the hippocampus of the brain (Couce et al. 1992), its expression is greatly reduced after birth (Sklar et al. 1992; Kornfeld 1992). Although the expression of the M6P/IGF-II receptor is reduced in adults, all cells nevertheless still express low levels of this receptor. The M6P/IGF-II receptor is present both in the cytoplasmic and on the surface of cells (Kornfeld 1992; Nissley and Lopaczynski 1991). It functions primarily in the sorting and trafficking of newly synthesized lysosomal enzymes to the lysosomes. It has now, however, also been shown to bind the growth factors, cathepsin D, proliferin, and the latent complex of TGF-β (Kornfeld 1992; Nissley and Lopaczynski 1991; Kovacina et al. 1989). The binding of the latent complex of TGF-β to the M6P/IGF-II receptor results in its internalization into the lysosomal compartment where it is degraded. It has, however, also been shown that the binding of the TGF-β latent complex to the M6P/IGF-II receptor significantly facilitates the activation of TGF-β by the proteolytic enzyme plasmin (Dennis and Rifkin 1991) when in the presence of transglutaminase (Kojima et al. 1993).

To determine whether the increased level of TGF-β observed during PB-treatment may result from an increase in the M6P/IGF-II receptor concentration in hepatocytes, we also immunohistochemically stained for this receptor. Normal tissue stained very weakly for the M6P/IGF-II receptor, and the pericentral hepatocytes had a slightly higher concentration of this receptor than those cells in the periportal region of the liver lobule (Jirtle et al. 1991). In contrast, the periportal hepatocytes stained intensely for the M6P/IGF-II receptor in animals that were PB-treated for 2 months (unpublished results). The elevated expression of the M6P/IGF-II receptor colocalized in the cells that contained increased levels of TGF-β. This suggests that the PB-induced increase in the intracellular concentration of TGF-β results from an enhanced uptake of the TGF-β latent complex upon binding to the M6P/IGF-II receptor. The PB-induced increase in the M6P/IGF-II receptor expression in hepatocytes appears to be due in part to an increase in the steady state mRNA levels. The molecular mechanism by which PB increases M6P/IGF-II receptor expression is unknown but is presently being investigated.

Thus, we have shown that PB causes an increase in the concentration of the M6P/IGF-II receptors in normal hepatocytes. Therefore, our results suggest that the elevated level of TGF-β in PB-treated hepatocytes likely results from an enhanced uptake because of the increase in the concentration of hepatic M6P/IGF-II receptors. These results further suggest that PB-treated hepatocytes are also exposed to a high concentration of active TGF-β, and that this results in the observed reduction in their ability to readily respond to mitogenic stimuli.

In contrast to the elevated expression of TGF-β found in PB-treated hepatocytes, the initiated cells in a portion of the PB-promoted hepatic nodules contain significantly less TGF-β (Jirtle and Meyer 1991). Interestingly, these nodules also lack the ability to upregulate the M6P/IGF-II receptor gene in response to PB (unpublished results). This suggests that the initiated cells in this subset of nodules have a growth advantage relative to the surrounding normal hepatocytes because they are unable to readily activate the potent growth inhibitor TGF-β. Additionally, these tumor cells would be deficient in their ability to degrade IGF-II, a growth factor often overexpressed in a variety of tumors, including those in the breast and liver (Daughaday 1990; Osborne et al. 1989; Yang and Rogler 1991). Thus, the inability of these initiated hepatocytes to upregulate the M6P/IGF-II receptor results in the initiated cells having a growth advantage over normal hepatocytes because of their relative inability (a) to activate the growth inhibitor TGF-β , and (b) to inactivate a potent growth factor IGF-II. Though only a small fraction of the nodules at the early stage of development lack the ability to upregulate the M6P/IGF-II receptor in response to PB, this altered phenotype becomes increasingly more prevalent as the tumors progress to malignant carcinomas.

Interestingly, the loss of M6P/IGF-II receptor regulation appears to be a relatively common event in liver carcinogenesis, especially during the latter stages of carcinoma development. This may result from the fact that this gene is imprinted (Barlow et al. 1991; Stöger et al. 1993). Consequently, only a single allele needs to be inactivated in order to observe a loss of gene function. Since the putative imprinting signal consists of a single methylated cytosine in an intron 27 kb upstream from the initiation start site, potentially the loss of only a single methyl group could inactivate the function of this gene (Stöger et al. 1993).

In conclusion, we have demonstrated that long-term PB exposure causes a substantial reduction in the proliferative capacity of normal hepatocytes. This is correlated with a significant elevation in the level of TGF-β in the liver. Furthermore, concomitant with this increase in TGF-β is a dramatic elevation in the cellular expression of the M6P/IGF-II receptor, a receptor that has been shown to facilitate the activation of TGF-β. We are suggesting that tumors unable to upregulate the M6P/IGF-II receptor can not readily activate TGF-β. Thus, they would have a selective growth advantage over the surrounding normal hepatocytes. These data suggest that liver tumor promotion with PB is a process of natural selection for cells resistant to the growth inhibitory effect produced by PB, a liver tumor promotion model initially proposed by Solt and Farber (1976).

9.2 Mammary Tumor Chemoprevention

The monoterpene, d-limonene, which is found in the peel of oranges has been shown to both prevent the formation of chemical carcinogen-initiated rat mammary tumors (Elegbede et al. 1984; Haag et al. 1992) and cause the complete regression of large established mammary tumors (Jirtle et al. 1993). Consequently, limonene has been approved for phase I clinical trials in the UK and is presently under consideration for phase I trials in the US. Though the monoterpenes are effective chemopreventive and cytostatic agents, the mechanisms by which they produce these effects are presently unclear. Limonene has been shown to block the isoprenylation of G proteins, including *ras* (Crowell et al. 1991). Thus, limonene may produce these important anticarcinogenic effects in part by blocking signal transduction.

Recently, it has been suggested that another cytostatic agent, tamoxifen, functions indirectly by inducing an increased release of the potent mitoinhibitory agent TGF-β (Butta et al. 1992). Because of this interesting observation, we investigated whether limonene also caused an increase in the expression of TGF-β during mammary tumor regression. We also determined whether the concentration of the M6P/IGF-II receptor was increased in limonene-induced regressing mammary tumors since this receptor is known to play a central role in the activation of TGF-β (Dennis and Rifkin 1991). The expression of both TGF-β1 and the M6P/IGF-II receptor were slightly elevated in mammary adenocarcinoma cells when compared to that observed in normal mammary gland epithelium (Jirtle et al. 1993). The expression of TGF-β 1 and the M6P/IGF-II receptor were, however, dramatically increased in both the early and late stages of limonene-induced mammary tumor regression (Jirtle et al. 1993). This suggests that limonene causes the regression of mammary tumors in part by increasing the expression of TGF-β.

As previously discussed, however, TGF-β is not secreted as an active molecule but rather as a latent complex. The finding that the level of both TGF-β and the M6P/IGF-II receptor are increased concomitantly during limonene-induced regression of mammary tumors suggests not only that the TGF-β1 level is increased but that it is also efficiently activated because of the increase in the M6P/IGF-II receptors. Furthermore, the mRNA level demonstrated that the expression of the M6P/IGF-II receptor in the limonene-induced regressing tumors was

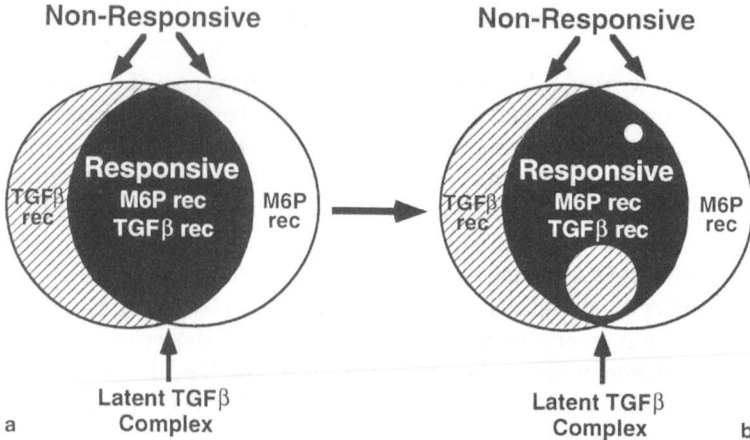

Fig. 4a,b. Model for mammary tumor responsiveness to *d*-limonene. Initially (**a**), approximately 80% of mammary tumors respond to limonene (*black area*). The nonresponding tumors are postulated to arise because they either lack a functional transforming growth factor (*TGF-β*) receptor signaling pathway (*gray area*) or have lost the ability to express the mannose 6-phosphate/insulin-like growth factor II (M6P/IGF-II) receptor (i.e., activate TGF-β; *hatched area*). The tumors that initially were responsive become nonresponsive because they either lose the function of the TGF-β receptor signaling pathway (*small gray circle*) or lose the ability to express the M6P/IGF-II receptor (i.e., activate TGF-β) (*hatched circle*)

significantly increased above that found in untreated tumors. In contrast, the tumors unresponsive to limonene did not have an increased expression in the M6P/IGF-II receptor mRNA level. These results are consistent with the hypothesis that limonene cannot induce the regression of mammary tumors if they are unable to upregulate the M6P/IGF-II receptor because these tumors do not have the capacity to activate TGF-β.

This postulate is not only interesting from the standpoint of mammary tumor progression, but also suggests a mechanism by which mammary tumors that are initially responsive to cytostatic agents ultimately are able to escape from their mitoinhibitory/differentiation effects. Our working model by which mammary tumors escape from the mitoinhibitory effect of cytostatic agents is shown on Fig. 4. In this model, we have focused only on the role of TGF-β in inhibiting tumor growth in response to cytostatic agents. Tumors are initially responsive if they contain both the TGF-β receptors involved in the signal transduction and the M6P/IGF-II receptor required for the activation of TGF-β. If the tumors lack either the TGF-β signaling receptor(s) and/or are unable to upregulate the M6P/IGF-II receptor in response to a cytostatic agent, the tumors will initially be unresponsive. The unresponsive tumors will continue to grow in the presence of a cytostatic agent, whereas the responsive tumors will regress. A portion of the responsive tumors will, however, ultimately escape from the growth inhibitory effect of the cytostatic agent. Again, we are suggesting that an "escaping tumor" can grow in the presence of the cytostatic agent because it either lost the ability to transduce a signal through the TGF-β receptors or alternatively has lost the ability to upregulate the M6P/IGF-II receptor and therefore enhance the activation of TGF-β. The probability that the latter scenario is more likely than the first stems from the important fact that the gene coding for the M6P/IGF-II receptor is imprinted. Thus, only a single allele needs to be genetically altered in order to lose gene function.

9.3 Summary

Both d-limonene and PB are mitoinhibitory agents that ultimately promote the formation of tumors that are resistant to the growth inhibitory effects of these agents. One cellular alteration that appears to be mechanistically involved with the limonene-resistant tumor phenotype is a loss in the ability to upregulate the expression of M6P/IGF-II receptor. This is also the same loss of gene function that we have implicated as playing an important role in liver tumor promotion. The apparent susceptibility of the M6P/IGF-II gene to loss of function may result from the gene being imprinted. Additionally, because of the M6P/IGF-II receptor's ability to facilitate the activation of TGF-β, a potent growth inhibitor, the gene that codes for this receptor functions as a tumor suppressor. In conclusion, liver tumor promotion and breast cancer chemoprevention appear to be mechanistically linked because both of these biological phenomena are, in part, dependent upon the growth inhibitory effects of TGF-β and the involvement of the M6P/IGF-II receptor in its activation.

Acknowledgments. We extend our thanks to Ms. Roxanne Scroggs for typing this manuscript. This work was supported by USPHS grants CA25951 and CA40172.

References

Abanobi SE, Lombardi B, Shinozuka H (1982) Stimulation of DNA synthesis and cell proliferation in the liver of rats fed a choline-devoid diet and their suppression by phenobarbital. Cancer Res 42:412–415

Barbason H, Rassenfosse C, Betz EH (1983) Promotion mechanism of phenobarbital and partial hepatectomy in DENA hepatocarcinogenesis cell kinetics effect. Br J Cancer 47:517–525

Barlow DP, Stöger R, Herrmann BG, Saito K, Schweifer N (1991) The mouse insulin-like growth factor type-2 receptor is imprinted and closely linked to the Tme locus. Nature 349:84–87

Border WA, Ruoslahti E (1992) Transforming growth factor-β in disease: the dark side of tissue repair. J Clin Invest 90:1–7

Brockenbrough JS, Meyer SA, Li C, Jirtle RL (1991) A reversible and phorbol ester-specific defect of protein kinase C translocation in hepatocytes isolated from phenobarbital-treated rats. Cancer Res 51:130–136

Butta A, MacLennan K, Flanders KC, Sacks NPM, Smith I, McKinna A, Dowsett M, Wakefield LM, Sporn MB, Baum N, Colletta AA (1992) Induction of transforming growth factor β1 in human breast cancer in vivo following tamoxifen treatment. Cancer Res 52:4261–4264

Chen R-H, Ebner R, Derynck R (1993) Inactivation of the type II receptor reveals two receptor pathways for the diverse TGF-β activities. Science 260:1335–1338

Couce ME, Weatherington AJ, McGinty JF (1992) Expression of insulin-like growth factor-II (IGF-II) and IGF-II/mannose-6-phosphate receptor in the rat hippocampus: an in situ hybridization and immunocytochemical study. Endocrinology 131:1636–1642

Crowell PL, Chang RR, Ren Z, Elson CE, Gould MN (1991) Selective inhibition of isoprenylation of 21–26 kDa proteins by the anticarcinogen d-limonene and its metabolites. J Biol Chem 266:17679–17685

Daughaday WH (1990) The possible autocrine/paracrine and endocrine roles of insulin-like growth factors of human tumors. Endocrinology 127:1–4 (editorial)

Dennis PA, Rifkin DB (1991) Cellular activation of latent transforming growth factor β requires binding to the cation-independent mannose 6-phosphate/insulin-like growth factor type II receptor. Proc Natl Acad Sci USA 88:580–584

Dragan YP, Pitot HC (1992) The role of the stages of initiation and promotion in phenotypic diversity during hepatocarcinogenesis. Carcinogenesis 5:739–750

Ebner R, Chen R-H, Shum L, Lawler S, Zioncheck TF, Lee A, Lopez AR, Derynck R (1993) Cloning of a type I TGF-β receptor and its effect on TGF-β binding to the type II receptor. Science 260:1344–1348

Elegbede JA, Elson CE, Qureshi A, Tanner MA, Gould MN (1984) Inhibition of DMBA-induced mammary cancer by the monoterpene d-limonene. Carcinogenesis 5:661–664

Haag JD, Lindstrom MJ, Gould MN (1992) Limonene-induced regression of mammary carcinomas. Cancer Res 52:4021–4026

Haig D, Graham C (1991) Genomic imprinting and the strange case of the insulin-like growth factor II receptor. Cell 64:1045–1046

Jirtle RL, Meyer SA (1991) Liver tumor promotion: effect of phenobarbital on EGF and protein kinase C signal transduction and transforming growth factor-β1 expression. Dig Dis Sci 36:659–668

Jirtle RL, Meyer SA, Brockenbrough JS (1991) Liver tumor promoter phenobarbital: a biphasic modulator of hepatocyte proliferation. In: Butterworth BE, Slaga TJ, Farland W, McClain M (eds) Chemically induced cell proliferation implications for risk assessment. Wiley-Liss, New York, pp 209–216

Jirtle RL, Haag JD, Ariazi EA, Gould MN (1993) Increased mannose 6-phosphate/insulin-like growth factor II receptor and TGF-β1 levels during monoterpene-induced regression of mammary tumors. Cancer Res 53:3894–3852

Jirtle RL, Hankins GR, Reisenbichler H, Boyer IJ (1994) Regulation of mannose 6-phosphate/insulin-like growth factor-II receptors and transforming growth factor beta during liver tumor promotion with phenobarbitol. Carcinogenesis 15: (in press)

Kankaki T, Olofsson A, Moren A, Wernstedt C, Hellman U, Miyazono K, Claesson-Welsh L, Heldin C-H (1990) TGF-β1 binding protein: a component of the large latent complex of TGF-β1 with multiple repeat sequences. Cell 61:1051–1061

Kojima S, Nara K, Rifkin DB (1993) Requirement for transglutaminase in the activation of latent transforming growth factor-β in bovine endothelial cells. J Cell Biol 121:439–448

Kornfeld S (1992) Structure and function of the mannose 6-phosphate/insulin like growth factor II receptors. Annu Rev Biochem 61:307–330

Kovacina KS, Steele-Perkins G, Purchio AF, Lioubin M, Miyazono K, Heldin C-H, Roth RA (1989) Interactions of recombinant and platelet transforming growth factor-β1 precursor with the insulin-like growth factor II/mannose 6-phosphate receptor. Biochem Biophys Res Commun 160:393–403

Laureys G, Barton DE, Ullrich A, Francke U (1988) Chromosomal mapping of the gene for the type II insulin-like growth factor receptor/cation-independent mannose 6-phosphate receptor in man and mouse. Genomics 3:224–229

Lin HY, Wang X-F, Hg-Eaton E, Weinberg RA, Lodish HF (1992) Expression cloning of the TGF-β type II receptor, a functional transmembrane serine/threonine kinase. Cell 68:775–785

Lopez-Cassillas F, Cheifetz S, Doody J, Andres JL, Lane WS, Massague J (1991) Structure and expression of the membrane proteoglycan betaglycan, a component of the TGF-β receptor system. Cell 67:785–795

Lopez-Cassillas F, Wrana JL, Massagué J (1993) Betaglycan presents ligand to the TGF-β signaling receptor. Cell 73:1435–1444

Massague J (1990) The transforming growth factor-β family. Annu Rev Cell Biol 6:597–641

McLean AE, Driver HE, Sutherland IA (1992) Liver tumour promotion by phenobarbital: comparison of rat and human studies. Prog Clin Biol Res 374:251–259

Nissley P, Lopaczynski W (1991) Insulin-like growth factor receptors. Growth Factors 5:29–43

Olofsson A, Miyazono K, Kanzaki T, Colosetti P, Engström U, Heldin C-H (1992) Transforming growth factor-β1, -β2, and -β3 secreted by a human glioblastoma cell line. Identification of small and different forms of large latent complexes. J Biol Chem 267:19482–19488

Osborne CK, Coronado EB, Kitten LJ, Arteaga CI, Fuqua SAW, Ramasharma K, Marshall M, Li CH (1989) Insulin-like growth factor II (IGF-II): a potential autocrine/paracrine growth factor for human breast cancer acting via the IGF-I receptor. Mol Endocrinol 3:1701–1709

Peraino C, Fry RJM, Staffeldt E (1971) Reduction and enhancement by phenobarbital of hepatocarcinogenesis induced in the rat by 2-acetylaminofluorene. Cancer Res 31:1506–1512

Peraino C, Fry RJM, Staffeldt E, Christopher JP (1975) Comparative enhancing effects of phenobarbital, amobarbital, diphenylhydantoin, and dichlorodiphenyltrichloroethane on 2-acetylaminofluorene-induced hepatic tumorigenesis in the rat. Cancer Res 35:2884–2890

Preat V, Lans M, de Gerlache J, Taper H, Roberfroid M (1987) Influence of the duration and the delay of administration of phenobarbital on its modulating effect on rat hepatocarcinogenesis. Carcinogenesis 8:333–335

Schlunegger MP, Grütter MG (1992) An unusual structural feature revealed by the 2.2 α resolution crystal structure of human transforming growth factor-β2. Nature 358:430–434

Schulte-Hermann R, Timmermann-Trosiener I, Schuppler J (1986) Facilitated expression of adaptive responses to phenobarbital in putative pre-stages of liver cancer. Carcinogenesis 7:1651–1655

Sklar MM, Thomas CL, Municchi G, Roberts CT Jr, LeRoith D, Kiess W, Nissley P (1992) Developmental expression of rat insulin-like growth factor-II/mannose 6-phosphate receptor messenger ribonucleic acid. Endocrinology 130:3484–3491

Solt D, Farber E (1976) New principle for the analysis of chemical carcinogenesis. Nature 263:701–703

Stöger R, Kubicka P, Liu C-G, Kafri T, Razin A, Cedar H, Barlow DP (1993) Maternal-specific methylation of the imprinted mouse Igf2r locus identifies the expressed locus as carrying the imprinting signal. Cell 73:61–71

Vogelstein B, Kinzler KW (1993) The multistep nature of cancer. Trends Genet 4:138–141

Wang X-F, Lin HY, Hg-Eaton E, Downward J, Lodish HF, Weinberg RA (1991) Expression cloning and characterization of the TGF-β type III receptor. Cell 67:797–805

Yager JD, Roebuck BD, Paluszcyk TL, Memoli VA (1986) Effects of ethinyl estradiol and tamoxifen on liver DNA turnover and new synthesis and appearance of gamma glutamyl transpeptidase-positive foci in female rats. Carcinogenesis 7:2007–2014

Yang D, Rogler CE (1991) Analysis of insulin-like growth factor II (IGF-II) expression in neoplastic nodules and hepatocellular carcinomas of woodchucks utilizing in situ hybridization and immunocytochemistry. Carcinogenesis 12:1893–1901

10 Peroxisome Proliferation and Hepatocarcinogenesis

B. G. Lake

10.1 Introduction

Peroxisomes (or "microbodies") are single membrane-limited cytoplasmic organelles present in the cells of animals, plants, fungi and protozoa. They are characterised by their content of catalase and a number of hydrogen peroxide-generating oxidase enzymes (Cohen and Grasso 1981; Reddy and Lalwani 1983). Like mitochondria, peroxisomes contain a complete fatty acid β-oxidation cycle (Lazarow and DeDuve 1976). In rat liver peroxisomes are normally spherical or oval in shape, approximately 0.5 μm in diameter and contain a finely granular matrix with a crystalline nucleoid core (Cohen and Grasso 1981).

Table 1. Some characteristics of peroxisome proliferation in rat and mouse liver

Liver weight
1. Liver enlargement due to both hepatocyte hyperplasia and hypertrophy
2. Increased replicative DNA synthesis (may be either transient or sustained)[a]

Morphological changes
1. Increased number and size of peroxisomes
2. Many "coreless" peroxisomes observed[b]
3. Increased smooth endoplasmic reticulum
4. Lysosomal changes and lipofuscin deposition[a]
5. Liver nodules and hepatocellular carcinoma[a]

Biochemical changes
1. Selective induction of peroxisomal enzymes (e.g. marked induction of peroxisomal fatty acid β-oxidation enzymes but only a small increase in catalase activity)
2. Induction of microsomal fatty acid (ω-1)- and particularly ω-oxidising enzyme activities (due to induction of cytochrome P-450 isoenzymes in the CYP4A subfamily)
3. Induction of carnitine acetyltransferase activity
4. Increase in an 80-kDa molecular weight polypeptide (due to induction of component enzymes of the peroxisomal fatty acid β-oxidation cycle)
5. Induction of cytosolic epoxide hydrolase
6. Inhibition of GSH peroxidase, GSH S-transferase and superoxide dismutase activities

For further details see Bentley et al. (1993); Cohen and Grasso (1981); Lake (1993); Lock et al. (1989); Moody et al. (1991); Reddy and Lalwani (1983) and Stott (1988).
GSH, glutathione.
[a]Depends on test compound, dose and duration of treatment.
[b]Normal rat and mouse liver peroxisomes contain a crystalline nucleoid core consisting of insolubilised urate oxidase.

10.2 Characteristics of Hepatic Peroxisome Proliferation in Rodents

A wide variety of chemicals have been shown to produce liver enlargement, peroxisome proliferation and induction of peroxisomal and microsomal fatty acid-oxidising enzyme activities in rodents (Bentley et al. 1993; Cohen and Grasso 1981; Lake and Lewis 1993; Lock et al. 1989; Moody et al. 1991; Reddy and Lalwani 1983; Stott 1988). Some characteristics of peroxisome proliferation in rat and mouse hepatocytes are shown in Table 1. Liver enlargement is due to both hyperplasia and hypertrophy, and organelle proliferation is associated with a differential induction of peroxisomal enzyme activities. While the enzymes of the peroxisomal fatty acid β-oxidation cycle, normally assessed as cyanide-insensitive palmitoyl-coenzyme A (CoA) oxidation (Lazarow and DeDuve 1976), are markedly induced, only small changes are observed in other peroxisomal enzymes such as catalase and D-amino acid oxidase. Peroxisome proliferators also markedly stimulate carnitine acetyltransferase and microsomal fatty acid (ω-1) and particularly ω-hydroxylase activities. The stimulation of microsomal fatty acid-oxidising enzymes, normally assessed as lauric acid 12-hydroxylase, is due to induction of cytochrome P-450 isoenzymes in the CYP4A subfamily (Gibson 1989; Sharma et al. 1988a,b). Carnitine acetyltransferase activity is found in peroxisomal, mitochondrial and microsomal fractions and, hence, induction of this enzyme may reflect induction in more than one subcellular compartment (Cohen and Grasso 1981; Bieber et al. 1981; Ishii et al. 1980; Moody et al. 1991). Lobular differences in the effects of peroxisome proliferators in rodent liver have also been observed. For example, in rat liver peroxisome proliferation – in contrast to cell replication – is more marked in centrilobular than in periportal hepatocytes (Barrass et al. 1993; Bell et al. 1991; Eacho et al. 1991; Eldridge et al. 1990). In rat liver correlations have been reported between the induction of peroxisomal fatty acid β-oxidation (palmitoyl-CoA oxidation) and organelle proliferation (Lin 1987; Sharma et al. 1988a). Other studies have demonstrated correlations between the induction of peroxisomal and microsomal fatty acid-oxidising enzyme activities (Dirven et al. 1992; Lake et al. 1984a; Sharma et al. 1988a,b). Studies in several laboratories have demonstrated that the characteristics of peroxisome proliferation in vivo may also be observed in vitro in primary hepatocyte cultures (Bieri 1993; Foxworthy and Eacho 1994; Lake and Lewis 1993).

10.3 Rodent Liver Peroxisome Proliferators

A very large number of chemicals have been shown to produce peroxisome proliferation in rat and mouse hepatocytes (Bentley et al. 1993; Cohen and Grasso 1981; Lake and Lewis 1993; Lock et al. 1989; Moody et al. 1991; Reddy and Lalwani 1983; Stott 1988). Some examples of the different classes of chemicals which produce this effect are shown in Table 2. Although peroxisome proliferators appear to be

Table 2. Examples of classes of chemicals which produce peroxisome proliferation in rodent liver

Chemical class	Examples
Therapeutic agents	Acetylsalicylic acid, bezafibrate, bifonazole, ciprofibrate, clofibrate, DL-040, fenofibrate, LY 171883, tiadenol, nafenopin, Wy-14,643
Steroids	Dehydroepiandrosterone
Herbicides	2,4-Dichlorophenoxyacetic acid, fomesafen, lactofen, 2,4,5-trichlorophenoxyacetic acid
Plasticizers	Di-(2-ethylhexyl)adipate (DEHA), di-(2-ethylhexyl)phthalate (DEHP), di-(2-ethylhexyl)terephthalate, di-(isodecyl)phthalate, di-(isononyl)phthalate, tri-(2-ethylhexyl)trimellitate
Solvents and industrial chemicals	Chlorinated paraffins, perchloroethylene, perfluoro-n-octanoic acid, trichloroethylene
Food flavours and natural products	Cinnamyl anthranilate, citral linalool

For further details see Bentley et al. (1993), Cohen and Grasso (1981), Lake and Lewis (1993), Lock et al. (1989), Moody et al. (1991), Reddy and Lalwani (1983) and Stott (1988).

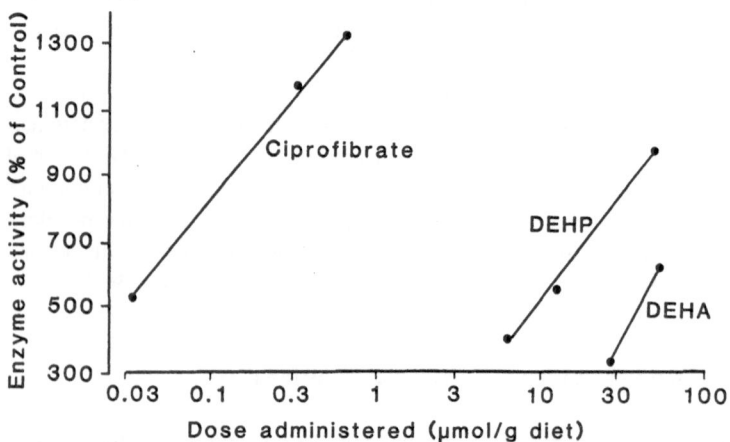

Fig. 1. Comparison of the potency of ciprofibrate, di-(2-ethylhexyl)phthalate (*DEHP*) and di-(2-ethylhexyl)adipate (*DEHA*) to produce peroxisome proliferation (assessed as induction of palmitoyl-coenzyme A oxidation) in rat liver. Male F344 rats were fed various dietary levels of the test compounds (see Fig. 2) for 30 days. The dose administered is expressed as µmol test compound/g diet and is plotted on a logarithmic scale. Data from Reddy et al. (1986)

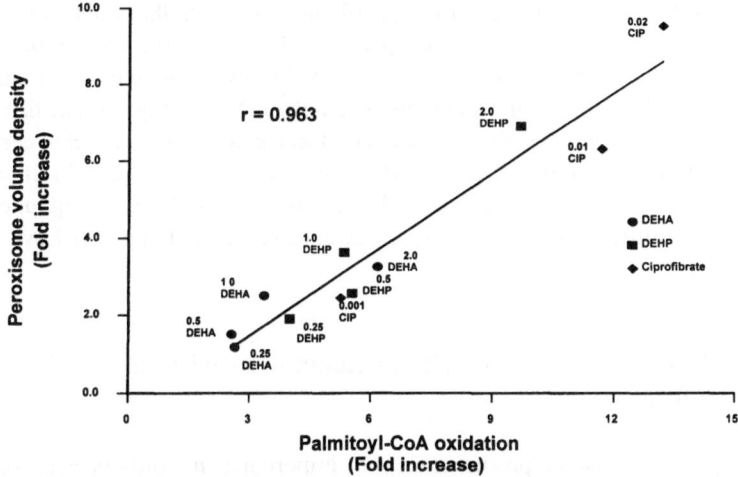

Fig. 2. Relationship between induction of hepatic palmitoyl-CoA oxidation and peroxisome volume density (determined by morphometric analysis) in male F344 rats fed diets containing 0.001%–0.02% ciprofibrate (CIP), 0.25%–2.0% di-(2-ethylhexyl)phthalate (*DEHP*) and 0.25%–2.0% di-(2-ethylhexyl)-adipate (*DEHA*) for 30 days. Data from Reddy et al. (1986)

structurally diverse, at least for some compounds, similarities in their three-dimensional structures have been reported (Lake and Lewis 1993; Lake et al. 1988). Data from both in vivo and in vitro studies has demonstrated clear structure–activity relationships for various classes of peroxisome proliferators (Lake and Lewis 1993; Lake et al. 1993).

A characteristic feature of many, but not all, peroxisome proliferators is the presence of acidic function (Lake and Lewis 1993; Lake et al. 1988; Lock et al. 1989). This acidic function is normally a free carboxyl group, either present as a free carboxyl group in the parent structure or one that is unmasked by metabolism. Alternatively, the compound may contain a chemical grouping which is a bioisostere of a carboxyl group (Thornber 1979), such as the tetrazole moiety of LY 171883 (Eacho et al. 1986) and the sulphonamide moiety of fomesafen (Lock et al. 1989).

Rodent liver peroxisome proliferators are known to exhibit marked compound potency differences. For example, Reddy et al. (1986) compared the hepatic effects of ciprofibrate, di-(2-ethylhexyl)phthalate (DEHP) and di-(2-ethylhexyl)adipate (DEHA) in the rat. Ciprofibrate was far more potent in inducing hepatic palmitoyl-CoA oxidation than DEHP, which was somewhat more potent than DEHA (Fig. 1). The data from this study also illustrates a good correlation (Fig. 2) between induction of palmitoyl-CoA oxidation and organelle proliferation (determined by morphometric analysis). Other examples of compound potency differences have been reviewed elsewhere (Lake and Lewis 1993).

10.4 Hepatocarcinogenicity of Rodent Peroxisome Proliferators

Hepatic peroxisome proliferation is of importance, not only because of the wide range of chemicals which produce this effect in rodents, but because a number of these agents have been shown to increase the incidence of liver tumours in rats and mice (Bentley et al. 1993; Cohen and Grasso 1981; Moody et al. 1991; Reddy and Lalwani 1983; Stott 1988). Some examples of peroxisome proliferators shown to be hepato-carcinogenic in rats and/or mice are shown in Table 3. It should be noted that peroxisome proliferators may also produce tumours in other organs, such as the pancreas and testis (Hinton and Price 1993).

Although peroxisome proliferators can produce hepatocellular carci-noma in rodents, they are not considered to be genotoxic agents. For example, studies with several peroxisome proliferators have shown that they do not bind covalently to hepatic DNA after in vivo administration to either rats or mice (Albro et al. 1983; Goel et al. 1985; Von Däniken et al. 1981, 1984). In addition, peroxisome proliferators have been extensively tested in a wide range of short-term tests for genotoxicity (Budroe and Williams 1993). Generally, peroxisome proliferators pro-duce negative results in such short-term tests, although a few positive findings have been reported (Budroe and Williams 1993; Hwang et al. 1993). In keeping with the properties of nongenotoxic carcinogens, peroxisome proliferators do not produce liver tumours when tested in initiation studies (Popp and Cattley 1993). However, when appropriate histochemical markers are employed, several peroxisome proliferators have been demonstrated to be effective in rat liver tumour promotion

Table 3. Examples of peroxisome proliferators shown to be hepatocarcinogenic in rodents

Compound	Susceptible species[a]	Reference
Bezafibrate	Rat	Reddy and Lalwani (1983)
BR-931	Rat, mouse	Reddy and Lalwani (1983)
Ciprofibrate	Rat	Rao et al. (1984)
Clofibrate	Rat	Reddy and Lalwani (1983)
DEHA	Mouse	NTP (1982a)
DEHP	Rat, mouse	NTP (1982b)
LY-171883	Mouse	Bendele et al. (1990)
Methylclofenapate	Rat	Reddy and Lalwani (1983)
Nafenopin	Rat, mouse	Reddy and Lalwani (1983)
Trichloroethylene	Mouse	NCI (1976)
Wy-14,643	Rat, mouse	Reddy and Lalwani (1983)

DEHA, di-(2-ethylhexyl)adipate; DEHP, di-(2-ethylhexyl)phthalate.
[a]Susceptible species indicated for where a significant increase in liver tumours was observed under the conditions of the particular bioassay. In some studies species and/or sex differences have been observed. For example, at the dose levels employed DEHA and trichloroethylene were found to produce liver tumours in the mouse but not in the rat (NCI 1976; NTP 1982a).

studies (Cattley and Popp 1989; Popp and Cattley 1993). Moreover, differences have been observed between the properties of peroxisome proliferators and other known liver tumour promoters, such as phenobarbitone (Cattley and Popp 1989).

10.5 Mechanisms of Peroxisome Proliferator-Induced Hepatocarcinogenesis

Several hypotheses have been proposed for the mechanism of peroxisome proliferator-induced hepatocarcinogenesis in rodents. These mechanisms include:

1. Induction of sustained oxidative stress (Reddy and Lalwani 1983; Reddy and Rao 1989)
2. A role for increased cell proliferation (Marsman et al. 1988; Popp and Marsman 1991)
3. The promotion of spontaneously formed preneoplastic liver lesions (Cattley et al. 1991; Grasl-Kraupp et al. 1993)

Some aspects of these various mechanisms are briefly reviewed below together with the effect of peroxisome proliferators on apoptosis in rodent liver.

10.5.1 Oxidative Stress

Reddy and coworkers have proposed that the chronic administration of peroxisome proliferators produces a sustained oxidative stress in rodent hepatocytes due to an imbalance in the production and degradation of hydrogen peroxide (Reddy and Lalwani 1983; Reddy and Rao 1989; Reddy et al. 1986). Organelle proliferation is associated with a differential effect on peroxisomal enzyme activities (Table 1). The first enzyme of the peroxisomal fatty acid β-oxidation cycle, acyl-CoA oxidase, produces hydrogen peroxide and, hence, the cyclic oxidation of a single fatty acid molecule can result in the production of several molecules of hydrogen peroxide (Lazarow and DeDuve 1976). As peroxisome proliferators markedly induce the peroxisomal fatty acid β-oxidation cycle, but produce only a small increase in catalase activity, any excess hydrogen peroxide not destroyed by peroxisomal catalase can diffuse through the peroxisomal membrane (Cohen and Grasso 1981; Reddy and Lalwani 1983) into the cytosol, where it will be a substrate for cytosolic selenium-dependent glutathione (GSH) peroxidase. However, the administration of peroxisome proliferators is often associated with a reduction in cytosolic GSH peroxidase activity and other enzymes, including superoxide dismutase and GSH S-transferase (Conway et al. 1989; Elliot and Elcombe 1987; Goel et al. 1986; Lake 1993; Lake et al. 1987). The reduction in GSH peroxidase and GSH S-transferase activities may result in a reduced ability to detoxify organic hydroperoxides, including products of lipid peroxidation.

The enzyme changes described above are postulated to result in increased intracellular levels of hydrogen peroxide which, either directly or via other reactive oxygen species (e.g. hydroxyl radical), can attack membranes and DNA (Reddy and Lalwani 1983; Reddy and Rao 1989). A number of experimental observations have provided support for the involvement of oxidative stress in the hepatotoxicity of peroxisome proliferators (Lake 1993; Reddy and Rao 1989). These include:

1. Increased peroxisomal hydrogen peroxide formation
2. Demonstration in in vitro systems that fatty acid metabolism by peroxisomal fractions can result in hydroxyl radical formation and damage to added DNA
3. Increased lipid peroxidation
4. Increased lipofuscin deposition
5. Changes in levels of antioxidants
6. Inhibition of tumour formation by antioxidants
7. Increased levels of 8-hydroxydeoxyguanosine in DNA

Some of these data are briefly reviewed below.

A number of studies have examined the effect of peroxisome proliferators on lipid peroxidation and lipofuscin deposition in rodent liver (Lake 1993). Lipid peroxidation has been assessed either as levels of conjugated dienes or as thiobarbituric acid reactive material (primarily malonaldehyde). In short-term studies (up to 4 weeks) peroxisome proliferators have been reported not to increase hepatic lipid peroxidation (Elliot and Elcombe 1987; Lake 1993). However, in some, but not all, long-term studies increases in hepatic lipid peroxidation have been observed (Conway et al. 1989; Goel et al. 1986; Lake 1993; Lake et al. 1987).

A characteristic feature of the chronic administration of a number of peroxisome proliferators is the deposition of lipofuscin in hepatocytes of treated animals (Conway et al. 1989; Goel et al. 1986; Lake et al. 1987; Marsman et al. 1992; Reddy and Lalwani 1983). Lipofuscin, which is sometimes referred to as the "wear and tear" or "ageing" pigment, is considered to be associated with free radical and lipid peroxidative reactions. This material is stored within the cell as lysosomal residual bodies, and lipofuscin deposition is associated with increased hepatic lysosomal enzyme activity (Conway et al. 1989). In chronically treated animals, in which liver nodules and tumours are present, lipofuscin deposition is only observed in nonnodular and nontumorous portions of the liver (Lake et al. 1987; Reddy and Lalwani 1983).

A number of studies have demonstrated that peroxisome proliferators may modulate levels of both water-soluble (e.g. GSH) and lipid-soluble (e.g. vitamin E) antioxidants in rodent liver (Lake 1993). In other investigations the effect of antioxidants on peroxisome proliferator-induced liver tumour formation has been studied. Reddy and coworkers have examined the effect of coadministration of antioxidants, namely ethoxyquin and 2(3)-tert-butyl-4-hydroxyanisole (BHA) on ciprofibrate-induced hepatocarcinogenesis (Lalwani et al. 1983; Rao et al. 1984). The treatment of rats with ciprofibrate alone for 60 weeks resulted in a 100% incidence of liver tumours (Rao et al. 1984). While liver tumours were also present in all rats coadministered ciprofibrate with either ethoxyquin or BHA, the total number and size of these tumours was reduced. The effect of varying hepatic vitamin E levels on peroxisome proliferator-induced hepatocarcinogenesis has also been investigated (Glauert et al. 1990; Lake et al. 1991). Glauert et al. (1990) examined the effect of feeding various dietary levels of vitamin E on ciprofibrate-induced tumour formation. Hepatic vitamin E levels varied from 5.7-fold in rats fed the lowest compared to the highest vitamin E-supplemented diet. After 21 months of treatment the incidence of liver tumours was greatest in rats fed the highest levels of vitamin E (Glauert et al. 1990). Lake et al. (1991) examined the effect of feeding either an adequate or a vitamin E- and selenium-deficient diet on nafenopin-induced tumour formation. Levels of hepatic vitamin E in rats fed the deficient diet were 3%–10% of those fed the adequate diet. However, after 77 weeks treatment the incidence of liver tumours was significantly greater in rats fed the adequate rather than the deficient diet (Lake et al. 1991).

Several studies have investigated whether peroxisome proliferators can damage rodent hepatic DNA. Generally, studies employing alkaline elution assays to detect DNA strand breaks in rat liver have produced negative results (Bentley et al. 1988; Elliott and Elcombe 1987; Tamura et al. 1991). The short-term administration of peroxisome proliferators to rats failed to result in DNA adducts using the sensitive ^{32}P post-labelling technique, whereas the chronic administration of ciprofibrate and Wy-14,643 for up to 18 months did produce adducts in hepatic DNA (Gupta et al. 1985; Randerath et al. 1991). Although the nature of such adducts awaits elucidation, possibly the chronic administration of peroxisome proliferators and other nongenotoxic carcinogens may produce genetic alterations that can be detected by DNA adduct analysis (Randerath et al. 1991).

Oxygen radical attack on DNA is known to result in a variety of modified DNA bases (Imlay and Linn 1988) including 8-hydroxydeo-

Table 4. Effect of peroxisome proliferators on levels of 8-hydroxydeoxyguanosine in rat hepatic DNA

Compound	Treatment	Hepatic 8-OH-dG[a] (% control)	Reference
Ciprofibate	0.025% diet		
	16 weeks	140	Kasai et al. (1989)
	28 weeks	180	
	40 weeks	210	
DEHA	2.5% diet		
	1 week	155	Takagi et al. (1990a)
	2 weeks	145	
DEHP	1.2% diet		
	1 week	145	Cattley and Glover (1993)
	2 weeks	165	Takagi et al. (1990a,b)
	1 month	195	
	11 weeks	145	
	22 weeks	130	
	12 months	140	
Wy-14,643	0.005% diet		
	11 weeks	105	Cattley and Glover (1993)
	22 weeks	205	
	0.1% diet		
	11 weeks	150	
	22 weeks	190	

8-OH-dG, 8-hydroxydeoxyguanosine; DEHA, di-(2-ethylhexyl)adipate; DEHP, di-(2-ethylhexyl)phthalate.
[a]All studies performed in male F344 rats. Levels of 8-OH-dG in hepatic DNA from control animals ranged from 1.4–4.0 8-OH-dG/105dG.

xyguanosine (8-OH-dG). Several studies have examined the effect of treatment with peroxisome proliferators on levels of 8-OH-dG in rat hepatic DNA (Table 4). Although peroxisome proliferators do produce 1.4- to 2.1-fold increases in levels of 8-OH-dG, such changes do not appear to correlate with compound potency to produce tumours. For example, at bioassay doses both DEHP and DEHA produce similar increases in hepatic 8-OH-dG levels (Table 4), yet only DEHP produced liver tumours in male F344 rats (NTP 1982a,b).

10.5.2 Cell Replication

Many studies have demonstrated that cell proliferation is an important factor in the development of tumours by both genotoxic and nongenotoxic agents (Cohen and Ellwein 1990, 1991). For example, an enhanced rate of cell replication can increase the frequency of spontaneous lesions and the probability of converting DNA adducts from both endogenous and exogenous sources into mutations before they can be repaired (Cohen and Ellwein 1990, 1991; Popp and Marsman 1991).

Several peroxisome proliferators have been shown to produce a burst of cell replication during the first few days of treatment (Eacho et al. 1991; Moody et al. 1991; Reddy and Lalwani 1983). In addition, peroxisome proliferators may produce a sustained stimulation of replicative DNA synthesis in rodent liver. Popp and coworkers compared the effects of DEHP and Wy-14,643 in the rat at dose levels which produced a similar magnitude of hepatic peroxisome proliferation (Conway et al. 1989; Marsman et al. 1988). Both compounds produced a stimulation of cell replication in the first few days of treatment. However, when hepatocyte labelling index values were determined over a 7-day period, using subcutaneously implanted osmotic pumps to continuously infuse [^{3}H]thymidine, Wy-14,643 produced a 3.8- to 11.6-fold sustained increase in replicative DNA synthesis between 8 and 365 days of treatment (Marsman et al. 1988). In contrast, DEHP only produced a small increase after 365 days of treatment. A number of other studies have also examined whether the chronic administration of peroxisome proliferators can produce a sustained stimulation of replicative DNA synthesis in rodent hepatocytes. The data shown in Table 5 demonstrate that apart from intrinsic compound potency, dose is an important factor in determining whether a particular compound produces only a transient or a sustained stimulation of replicative DNA synthesis in rat hepatocytes. Thus low doses of nafenopin and Wy-14,643 do not produce sustained stimulation of cell replication, whereas higher doses do produce this effect.

It should be noted that peroxisome proliferators may exert lobular differences in effects on replicative DNA synthesis. For example, the short-term administration of Wy-14,643 was reported to produce a panlobular stimulation of cell replication in the mouse, but in the rat the increase was mainly confined to periportal hepatocytes (Eldridge et al.

Table 5. Effect of chronic administration of peroxisome proliferators on replicative DNA synthesis in rat hepatocytes

Compound	Treatment	Hepatocyte labelling index[a]	Reference
Ciprofibrate	0.025% diet 5 and 20 weeks	No increase	Yeldandi et al. (1989)
Clofibric acid	0.5% diet 5,11 and 22 weeks 26 weeks	No increase No increase	Marsman et al. (1992) Barrass et al. (1993)
DEHP	1.2% diet 39–151 days 365 days	No increase Increased	Marsman et al. (1988) Marsman et al. (1988)
Methyl-clofenapate	0.05% diet 26 weeks	Increased	Barrass et al. (1993)
Nafenopin	0.05% diet 30 days 15 and 40 weeks 0.1% diet 54 days	No increase No increase Increased	Eacho et al. (1991) Lake et al. (1993) Price et al. (1992)
Wy-14,643	0.0005% and 0.001% diet 3,6 and 13 weeks 0.005%, 0.01% and 0.1% diet 3,6 and 13 weeks 0.005% and 0.1% diet 5,11 and 22 weeks 0.025% diet 15 and 40 weeks	No increase Increased Increased Increased	Wada et al. (1992) Wada et al. (1992) Marsman et al. (1992) Lake et al. (1993)

DEHP, di-(2-ethylhexyl)phthalate.
[a]All studies performed in either F344 or Sprague-Dawley strain male rats. The DNA precursor (either [^{3}H]thymidine or 5-bromo-2,18-deoxyuridine) was administered continuously by a subcutaneously implanted osmotic pump for 3–7 days.

1990). A number of other peroxisome proliferators have also been reported to increase replicative DNA synthesis primarily in periportal hepatocytes in rat liver (Barrass et al. 1993; Eacho et al. 1991). Moreover, in chronic studies with Wy-14,643 (Marsman et al. 1988) and methylclofenapate (Barrass et al. 1993), the sustained stimulation of replicative DNA synthesis was also mainly confined to periportal hepatocytes.

10.5.3 Promotion of Spontaneous Preneoplastic Lesions

Several studies have demonstrated the presence of numerous foci of putative preneoplastic cells in the livers of untreated old rats and mice (Grasl-Kraupp et al. 1993). These lesions are considered to represent spontaneously initiated cells as they have biological characteristics similar to those of cells initiated by genotoxic carcinogens (Grasl-Kraupp et al. 1993). The ability of peroxisome proliferators to produce tumours in young compared to old rats has been investigated in studies with nafenopin (Kraupp-Grasl et al. 1991) and Wy-14,643 (Cattley et al. 1991). In both studies more adenomas and carcinomas were produced in old than in young rats.

10.5.4 Apoptosis

Several studies have demonstrated that nongenotoxic carcinogens can affect apoptosis (i.e. gene-directed cell death) in rodent liver (Bursch et al. 1992). Nafenopin has been reported to inhibit the apoptosis that occurs in rat liver after withdrawal of administration of the mitogenic agent cyproterone acetate (Bursch et al. 1986). However, in another study the administration of Wy-14,643 for 22 weeks resulted in an increase in apoptotic bodies in rat liver (Marsman et al. 1992). Transforming growth factor-$\beta 1$ (TGF-$\beta 1$) is known to induce apoptosis in rodent hepatocytes (Oberhammer et al. 1991) and the insulin-like growth factor II/mannose-6-phosphate (M6P/IGF-II) receptor is believed to be involved in transporting latent TGF-$\beta 1$ into hepatocytes (Jirtle et al. 1991). The short-term administration of methylclofenapate, nafenopin and Wy-14,643 has been shown to increase both TGF-$\beta 1$ and M6P/IGF-II receptor gene expression in rat liver (Rumsby et al. 1994). In another study, TGF-$\beta 1$ gene expression in rat liver was elevated after 1, 4 and 13 weeks treatment at dose levels of methylclofenapate, which resulted in a sustained stimulation of cell replication (Lake et al. 1994).

10.6 Species Differences in Hepatic Peroxisome Proliferation

Many studies have investigated species differences in hepatic peroxisome proliferation (Bentley et al. 1993; Cohen and Grasso 1981; Lock et al. 1989; Moody et al. 1991; Rodricks and Turnbull 1987; Stott 1988). Some representative studies are listed in Table 6. Clearly the rat and mouse may be considered responsive species to peroxisome proliferators. Based on both marker enzyme activities (e.g. palmitoyl-CoA oxidation, lauric acid 12-hydroxylase, carnitine acetyltransferase) and ultrastructural examination the Syrian hamster appears to exhibit an intermediate response, whereas in most studies the guinea pig is either nonresponsive or refractory (Table 6).

In assessing species differences in hepatic peroxisome proliferation a number of factors should be considered. These include the metabolism, disposition and dose of the test compound, sex differences, as well as intrahepatic differences in response. The importance of metabolism is illustrated by trichloroethylene which produces hepatic peroxisome proliferation in the mouse, while having little effect in the rat (Elcombe 1985). Metabolic studies demonstrated that trichloroethylene was extensively metabolised to trichloroacetic acid in the mouse, whereas this was a minor saturable route of metabolism in the rat. That the difference in trichloroacetic acid formation was responsible for the observed species difference was demonstrated by the fact that this compound produced peroxisome proliferation in rat and mouse hepatocytes both in vivo and in vitro (Elcombe 1985). With respect to compound disposition, DEHP is known to be more extensively absorbed after oral administration in the rat than in the marmoset (Rhodes et al. 1986). However, the observed in vivo species differences in response (Table 6) is supported by the observation that metabolites of DEHP which produce peroxisome proliferation in rat hepatocytes in vitro have no significant effect in cultured marmoset hepatocytes (Elcombe and Mitchell 1986). Generally, in vitro studies with primary hepatocyte cultures from the rat, mouse Syrian hamster, guinea pig and marmoset have supported the results of in vivo studies in these species (Bentley et al. 1993; Bieri 1993; Bieri et al. 1988; Elcombe 1985; Elcombe and Mitchell 1986; Lake et al. 1986).

Table 6. Some examples of species differences in hepatic peroxisome proliferation

Compound	Species examined[a]			References
	Responsive	Intermediate	Nonresponsive	
Bezafibrate	Rat, mouse	Syrian hamster, guinea pig	Dog, rabbit, rhesus monkey	Watanabe et al. (1989)
Ciprofibrate	Rat, mouse	Syrian hamster, rabbit	Guinea pig, marmoset	Graham et al. (1992), Makowska et al. (1992)
Clobuzarit	Rat, mouse	Syrian hamster	Dog, marmoset	Orton et al. (1984)
Clofibrate	Rat, mouse	Syrian hamster	Marmoset, rhesus monkey	Holloway et al. (1982), Lake et al. (1984b)
DEHP	Rat, mouse	Syrian hamster	Guinea pig, cynomolgus monkey, marmoset	Lake et al. (1984b), Osumi and Hashimoto (1978), Rhodes et al. (1986), Short et al. (1987)
Dehydroepiandrosterone	Rat, mouse	Syrian hamster	Guinea pig	Sakuma et al. (1992)
LY 171883	Rat, mouse	Syrian hamster	Guinea pig, dog, rhesus monkey	Eacho et al. (1986)
Nafenopin	Rat	Syrian hamster	Guinea pig, marmoset	Lake et al. (1989)

DEHP, di-(2-ethylhexyl)phthalate

[a] Peroxisome proliferation assessed by ultrastructural examination and/or measurement of marker enzyme activities. Intermediate species are less responsive than the rat and mouse, whereas nonresponsive species are either refractory or exhibit only a small response at high dosage. For further details see individual references.

Several studies have examined the ability of rodent peroxisome proliferators to produce effects in primates and humans. With respect to primates, studies with a number of compounds in both New (e.g. marmoset) and Old (e.g. rhesus monkey) World monkeys have failed to provide any evidence of significant hepatic peroxisome proliferation (Table 6). However, two compounds, namely ciprofibrate (Reddy et al. 1984) and DL-040 (Lalwani et al. 1985), have been reported to produce hepatic peroxisome proliferation in cynomolgus and/or rhesus monkeys, albeit at high doses. In humans studies have been conducted in patients treated with several hypolipidaemic agents (all being rodent peroxisome proliferators) including ciprofibrate, clofibrate, fenofibrate and gemfibrozil (Bentley et al. 1993). While most studies have failed to detect any significant changes, clofibrate was reported to produce a small increase in the number of peroxisomes (Hanefeld et al. 1983) and ciprofibrate to produce a small increase in the proportion of the hepatocyte cytoplasm occupied by peroxisomes (cited in Bentley et al. 1993). However, owing to the large interindividual variation in peroxisome morphometrics observed in these studies, together with cell-to-cell variations and lobular variations, it is difficult to attach any clear biological significance to these findings (Bentley et al. 1993). Generally, peroxisome proliferators have not been reported to produce any significant effects on marker enzyme activities and/or peroxisomes in cultured primate and human hepatocytes (Bentley et al. 1993; Bieri 1993; Bieri et al. 1988; Elcombe 1985; Elcombe and Mitchell 1986; Elcombe and Styles 1989; Foxworthy and Eacho 1994).

Although several studies have demonstrated species differences in hepatic peroxisome proliferation based on measurement of marker enzyme activities and ultrastructural examination, comparatively few investigations have evaluated species differences in cell replication. While both nafenopin and Wy-14,643 are potent mitogens in rat liver, they do not appear to produce any significant stimulation of replicative DNA synthesis in Syrian hamster hepatocytes after either acute or chronic administration (Price et al. 1992; Lake et al. 1993). In keeping with their known in vivo properties, certain peroxisome proliferators have been shown to stimulate DNA synthesis in rat hepatocyte cultures (Bieri 1993; Elcombe and Styles 1989; Marsman et al. 1993). However, in other studies methylclofenapate was found not to increase DNA synthesis in guinea pig, marmoset and human hepatocytes (Elcombe

and Styles 1989), and nafenopin was also ineffective in human hepatocytes (Parzefall et al. 1991). Although nafenopin has been reported to
increase DNA synthesis in marmoset hepatocytes, this was dependent
on the particular culture conditions employed (Bieri et al. 1988).

10.7 Conclusions

The bulk of the available data suggests that rodent liver peroxisome
proliferators are nongenotoxic carcinogens (Budroe and Williams
1993). Thus in order to assess the hazard, if any, of these compounds for
humans, information is required on the mechanisms(s) of tumour formation in susceptible species and data on species differences in response.

A number of mechanisms for peroxisome proliferator-induced hepatocarcinogenesis in rodents have been proposed (Fig. 3). Some studies
have suggested that the magnitude of peroxisome proliferation provides
an indirect measure of oxidative stress to hepatocytes and that this
correlates with subsequent tumour formation. For example, potency
differences exist between ciprofibrate, DEHP and DEHA not only for
peroxisome proliferation (Figs. 1,2) but also for tumour formation
(Lake 1993; NTP 1982a,b; Rao et al. 1984; Reddy et al. 1986). However, other studies have suggested that the magnitude of peroxisome
proliferation does not always correlate with carcinogenic potency (Lake
et al. 1993; Marsman et al. 1988, 1992). Indeed certain peroxisome
proliferators, at doses which produce a sustained stimulation of cell
replication, produce tumours more rapidly than other compounds which
produce a similar magnitude of peroxisome proliferation (Lake 1993;
Marsman et al. 1988, 1992; Popp and Cattley 1993).

If the various hypotheses shown in Fig. 3 are combined, then a role
for increased cell replication in peroxisome proliferator-induced hepatocarcinogenicity is readily identified. For example, if hepatocytes are
transformed by either oxidative stress-induced damage or by alternative
mechanisms, such initiated cells may be promoted to liver tumours by
enhanced cell replication. Certainly peroxisome proliferators are effective promoters of certain populations of initiated cells (Cattley and Popp
1989; Grasl-Kraupp et al. 1993; Marsman and Popp 1994; Popp and
Cattley 1993). Indeed, recent studies suggest that peroxisome proliferators can influence rates of both cell replication and cell death in

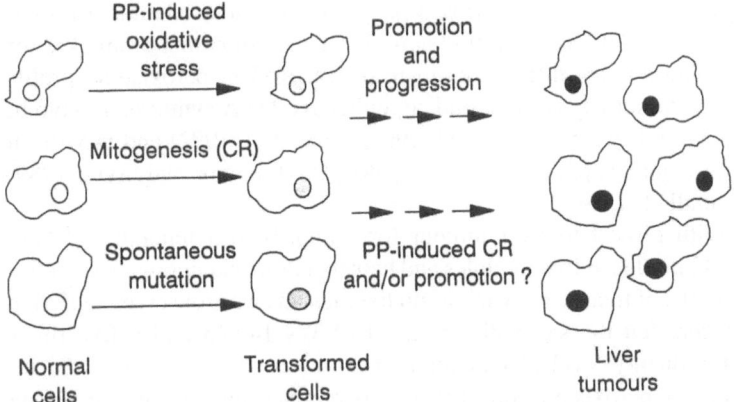

Fig. 3. Some possible mechanisms of peroxisome proliferator (*PP*)-induced liver tumour formation. Hepatocytes may become initiated by either oxidative stress-induced damage, enhanced cell replication (*CR*), or spontaneously. Growth selection of initiated cells and their promotion and progression to hepatocellular carcinoma may involve enhanced cell replication and inhibition of apoptosis

particular populations of hepatocytes. For example, basophilic nodules appear to be the precursors of peroxisome proliferator-induced tumours and such foci have high rates of replicative DNA synthesis (Marsman and Popp 1994). In addition, peroxisome proliferators also affect apoptosis and TGF-β1 gene expression in rodent liver (Bursch et al. 1986; Lake et al. 1994; Marsman et al. 1992; Rumsby et al. 1994). Further studies should help elucidate the precise mechanism(s) of peroxisome proliferator-induced tumour formation in susceptible species (i.e. the rat and mouse).

With respect to species differences in response, rats and mice are clearly responsive species, whereas the majority of both in vivo and in vitro studies suggest that primates, including humans, are either much less responsive or essentially refractory to rodent peroxisome proliferators. However, while effects on peroxisome morphology and marker enzyme activities have been extensively studied, few studies have examined species differences in peroxisome proliferator-induced cell replication and liver tumour formation. As enhanced cell replication

appears to play a role in peroxisome proliferator-induced tumour formation in rats and mice, it would appear to be an important biomarker for
assessing species differences in response. Rodent peroxisome proliferators do not appear to stimulate replicative DNA synthesis in vivo in
the Syrian hamster (Lake et al. 1993; Price et al. 1992) and in vitro in
cultured guinea pig and human hepatocytes (Elcombe and Styles 1989;
Parzefall et al. 1991).

With respect to liver tumour formation, both nafenopin and Wy-
14,643 produced liver nodules and hepatocellular carcinoma in rats after
60 weeks of treatment, whereas no liver lesions were observed in Syrian
hamsters fed the same dietary level of Wy-14,643 and a five times
higher dietary level of nafenopin (Lake et al. 1993). Possibly the inability of peroxisome proliferators to produce tumours in the Syrian
hamster may be associated with the fact that such compounds do not
appear to be mitogenic agents in this species. Two studies have provided
evidence that peroxisome proliferators may not produce liver lesions in
primates. In one ongoing study ciprofibrate was found not to produce
any morphological changes in marmoset liver after 3 years of administration (Graham et al. 1994). Similarly, clofibrate was found not to
increase liver weight or to produce liver tumours in marmosets after
6.5 years of administration (Tucker and Orton 1993). Clearly, further
carcinogenicity studies in partially responsive (e.g. Syrian hamster) and
nonresponsive (e.g. guinea pig) species would strengthen the conclusion
that peroxisome proliferators do not constitute any significant hazard for
humans.

Acknowledgement. We are grateful to the UK Ministry of Agriculture, Fish-
eries and Food for support of BIBRA studies on hepatic peroxisome proliferation.

References

Albro PW, Corbett JT, Schroeder JL, Jordan, ST (1983) Incorporation of
   radioactivity from labelled di-(2-ethylhexyl)phthalate into DNA of rat liver
   in vivo. Chem Biol Interact 44:1–16
Barrass NC, Price RJ, Lake BG, Orton TC (1993) Comparison of the acute and
   chronic mitogenic effects of the peroxisome proliferators methylclofenapate and clofibric acid in rat liver. Carcinogenesis 14:1451–1456

Bell DR, Bars RG, Gibson GG, Elcombe CR (1991) Localization and differential induction of cytochrome P450IVA and acyl-CoA oxidase in rat liver. Biochem J 275:247–252

Bendele AM, Hoover DM, van Lier RBL, Foxworthy PS, Eacho PI (1990) Effects of chronic treatment with the leukotriene D4-antagonist compound LY171883 on B6C3F1 mice. Fund Appl Toxicol 15:676–682

Bentley P, Bieri F, Muakkassah-Kelly S, Stäubli W, Waechter F (1988) Mechanisms of tumour induction by peroxisome proliferators. Arch Toxicol [Suppl] 12:240–247

Bentley P, Calder I, Elcombe C, Grasso P, Stringer D, Wiegand H-J (1993) Hepatic peroxisome proliferation in rodents and its significance for humans. Food Chem Toxicol 31:857–907

Bieber LL, Krahling JB, Clarke PRH, Valkner KJ, Tolbert NE (1981) Carnitine acyltransferases in rat liver peroxisomes. Arch Biochem Biophys 211:599–604

Bieri F (1993) Cultured hepatocytes: a useful in vitro system to investigate effects induced by peroxisome proliferators and their species specificity. In: Gibson G, Lake B (eds) Peroxisomes: biology and importance in toxicology and medicine. Taylor and Francis, London, pp 299–311

Bieri F, Stäubli W, Waechter F, Muakkassah-Kelly S, Bentley P (1988) Stimulation of DNA synthesis but not of peroxisomal β-oxidation by nafenopin in primary cultures of marmoset hepatocytes. Cell Biol Int Rep 12:1077–1087

Budroe JD, Williams GM (1993) Genotoxicity studies of peroxisome proliferators. In: Gibson G, Lake B (eds) Peroxisomes: biology and important in toxicology and medicine. Taylor and Francis, London, pp 525–568

Bursch W, Düsterberg B, Schulte-Hermann R (1986) Growth, regression and cell death in rat liver as related to tissue levels of the hepatomitogen cyproterone acetate. Arch Toxicol 59:221–227

Bursch W, Oberhammer F, Schulte-Hermann R (1992) Cell death by apoptosis and its protective role against disease. Trends Pharmacol Sci 13:245–251

Cattley RS, Glover SE (1993) Elevated 8-hydroxydeoxyguanosine in hepatic DNA of rats following exposure to peroxisome proliferators: relationship to carcinogenesis and nuclear localization. Carcinogenesis 14:2495–2499

Cattley RC, Popp JA (1989) Differences between the promoting activities of the peroxisome proliferator WY-14,643 and phenobarbital in rat liver. Cancer Res 49:3246–3251

Cattley RC, Marsman DS, Popp JA (1991) Age-related susceptibility to the carcinogenic effect of the peroxisome proliferator WY-14,643 in rat liver. Carcinogenesis 12:469–473

Cohen SM, Ellwein LB (1990) Cell proliferation in carcinogenesis. Science 249:1007–1011

Cohen SM, Ellwein LB (1991) Genetic errors, cell proliferation, and carcinogenesis. Cancer Res 51:6493–6505

Cohen AJ, Grasso P (1981) Review of the hepatic response to hypolipidaemic drugs in rodents and assessment of its toxicological significance to man. Food Chem Toxicol 19:585–605

Conway JG, Tomaszewski KE, Olson MJ, Cattley RC, Marsman DS, Popp JA (1989) Relationship of oxidative damage to the hepatocarcinogenicity of the peroxisome proliferators di(2-ethylhexyl)phthalate and Wy-14,643. Carcinogenesis 10:513–519

Dirven HAAM, van den Broek PHH, Peters JGP, Noordhoek J, Jongeneelen FJ (1992) Microsomal lauric acid hydroxylase activities after treatment of rats with three classical cytochrome P-450 inducers and peroxisome proliferating compounds. Biochem Pharmacol 43:2621–2629

Eacho PI, Foxworthy PS, Johnson WD, Hoover DM, White SL (1986) Hepatic peroxisomal changes induced by a tetrazole-substituted alkoxyacetophenone in rats and comparison with other species. Toxicol Appl Pharmacol 83:430–437

Eacho PI, Lanier TL, Brodhecker CA (1991) Hepatocellular DNA synthesis in rats given peroxisome proliferating agents: comparison of Wy-14,643 to clofibric acid, nafenopin and LY 171883. Carcinogenesis 12:1557–1561

Elcombe CR (1985) Species differences in carcinogenicity and peroxisome proliferation due to trichloroethylene: a biochemical human hazard assessment. Arch Toxicol [Suppl] 8:6–17

Elcombe CR, Mitchell AM (1986) Peroxisome proliferation due to di(2-ethylhexyl)phthalate (DEHP): species differences and possible mechanisms. Environ Health Perspect 70:211–219

Elcombe CR, Styles JA (1989) Species differences in peroxisome proliferation. Toxicologist 9:63

Eldridge SR, Tilbury LF, Goldsworthy TL, Butterworth BE (1990) Measurement of chemically induced cell proliferation in rodent liver and kidney: a comparison of 5-bromo-2'-deoxyuridine and [^{3}H]thymidine administered by injection or osmotic pump. Carcinogenesis 11:2245–2251

Elliott BM, Elcombe CR (1987) Lack of DNA damage or lipid peroxidation measured in vivo following treatment with peroxisomal proliferators. Carcinogenesis 8:1213–1218

Foxworthy PS, Eacho PI (1994) Cultured hepatocytes for studies of peroxisome proliferation: methods and applications. J Pharmacol Toxicol Methods 31:21–30

Gibson GG (1989) Comparative aspects of the mammalian cytochrome P450IV gene family. Xenobiotica 19:1123–1148

Glauert HP, Beaty MM, Clark TD, Greenwell WS, Tatum V, Chen L-C, Borges T, Clark TL, Srinivasan SR, Chow CK (1990). Effect of dietary vitamin E on the development of altered hepatic foci and hepatic tumors induced by the peroxisome proliferator ciprofibrate. J Cancer Res Clin Oncol 116:351–356

Goel SK, Lalwani ND, Fahl WE, Reddy JK (1985) Lack of covalent binding of peroxisome proliferators nafenopin and WY-14,643 to DNA in vivo and in vitro. Toxicol Lett 24:37–43

Goel SK, Lalwani ND, Reddy JK (1986) Peroxisome proliferation and lipid peroxidation in rat liver. Cancer Res 46:1324–1330

Graham MJ, Wilson SA, Winham MA, Spencer AJ, Rees JA, Old SL, Bonner FW (1994) Lack of peroxisome proliferation in marmoset liver following treatment with ciprofibrate for 3 years. Fund Appl Toxicol 22:58–64

Grasl-Kraupp B, Huber W, Schulte-Hermann R (1993) Are peroxisome proliferators tumour promoters in rat liver? In: Gibson G, Lake B (eds) Peroxisomes: biology and importance in toxicology and medicine. Taylor and Francis, London, pp 667–693

Gupta RC, Goel SK, Earley K, Singh B, Reddy JK (1985) ^{32}P-Postlabelling analysis of peroxisome proliferator-DNA adduct formation in rat liver in vivo and hepatocytes in vitro. Carcinogenesis 6:933–936

Hanefeld M, Kemmer C, Kadner E (1983) Relationship between morphological changes and lipid-lowering action of p-chlorophenoxyisobutyric acid (CPIB) on hepatic mitochondria and peroxisomes in man. Atheosclerosis 46:239–246

Hinton RH, Price SC (1993) Extrahepatic peroxisome proliferation and the extrahepatic effects of peroxisome proliferators. In: Gibson G, Lake B (eds) Peroxisomes: biology and importance in toxicology and medicine. Taylor and Francis, London, pp 487–511

Holloway BR, Thorp JM, Smith GD, Peter TJ (1982) Analytical subcellular fractionation and enzymatic analysis of liver homogenates from control and clofibrate-treated rats, mice and monkeys with reference to the fatty acid oxidizing enzymes. Ann NY Acad Sci 386:453–455

Hwang J-J, Hsai MTS, Jirtle RL (1993) Induction of sister chromatid exchange and micronuclei in primary cultures of rat and human hepatocytes by the peroxisome proliferator, Wy-14,643. Mutat Res 286:123–133

Imlay JA, Linn S (1988) DNA damage and oxygen radical toxicity. Science 240:1302–1309

Ishii H, Fukumori N, Horie S, Suga T (1980) Effects of fat content in the diet on hepatic peroxisomes of the rat. Biochim Biophys Acta 617:1–11

Jirtle RL, Carr BI, Scott CD (1991) Modulation of insulin-like growth factor-II/mannose 6-phosphate receptors and transforming growth factor-β1 during liver regeneration. J Biol Chem 266:22444–22450

Kasai H, Okada Y, Nishimura S, Rao MS, Reddy JK (1989) Formation of 8-hydroxydeoxyguanosine in liver DNA of rats following long-term exposure to a peroxisome proliferator. Cancer Res 49:2603–2605

Kraupp-Grasl B, Huber W, Taper H, Schulte-Hermann R (1991) Increased susceptibility of aged rats to hepatocarcinogenesis by the peroxisome proliferator nafenopin and the possible involvement of altered liver foci occurring spontaneously. Cancer Res 51:666–671

Lake BG (1993) Role of oxidative stress and enhanced cell replication in the hepatocarcinogenicity of peroxisome proliferators. In: Gibson G, Lake B (eds) Peroxisomes: biology and importance in toxicology and medicine, Taylor and Francis, London, pp 595–618

Lake BG, Lewis DFV (1993) Structure-activity relationships for chemically induced peroxisome proliferation in mammalian liver. In: Gibson G, Lake B (eds) Peroxisomes: biology and importance in toxicology and medicine. Taylor and Francis, London, pp 313–342

Lake BG, Gray TJB, Pels Rijcken WR, Beamand JA, Gangolli SD (1984a) The effect of hypolipidaemic agents on peroxisomal β-oxidation and mixed-function oxidase activities in primary cultures of rat hepatocytes. Relationship between induction of palmitoyl-CoA oxidation and lauric acid hydroxylation. Xenobiotica 14:269–276

Lake BG, Gray TJB, Foster JR, Stubberfield CR, Gangolli SD (1984b) Comparative studies on di-(2-ethylhexyl)phthalate-induced hepatic peroxisome proliferation in the rat and hamster. Toxicol Appl Pharmacol 72:46–60

Lake BG, Gray TJB, Gangolli SD (1986). Hepatic effects of phthalate esters and related compounds – in vivo and in vitro correlations. Environ Health Perspect 67:283–290

Lake BG, Kozlen SL, Evans JG, Gray TJB, Young PJ, Gangolli SD (1987) Effect of prolonged administration of clofibric acid and di-(2-ethylhexyl)phthalate on hepatic enzyme activities and lipid peroxidation in the rat. Toxicology 44:213–228

Lake BG, Lewis DFV, Gray TJB (1988) Structure-activity relationships for peroxisome proliferation. Arch Toxicol [Suppl] 12:217–224

Lake BG, Evans JG, Gray TJB, Körösi SA, North CJ (1989) Comparative studies on nafenopin-induced hepatic peroxisome proliferation in the rat, Syrian hamster, guinea pig, and marmoset. Toxicol Appl Pharmacol 99:148–160

Lake BG, Evans JG, Walters DG, Price RJ (1991) Comparison of the hepatic effect of nafenopin, a peroxisome proliferator, in rats fed adequate or vitamin E- and selenium-deficient diets. Hum Exp Toxicol 10:87–88

Lake BG, Evans JG, Cunninghame ME, Price RJ (1993) Comparison of the hepatic effects of nafenopin and Wy-14,643 on peroxisome proliferation and cell replication in the rat and Syrian hamster. Environ Health Perspect 101 [Suppl 5]:241–248

Lake BG, Rumsby PC, Cunninghame ME, Davies MJ, Price RJ (1994) Effect of methylclofenapate (MCP) on peroxisome proliferation, cell replication and transforming growth factor-β1 (TGF-β1) gene expression in rat liver. Toxicologist 14:301

Lalwani ND, Reddy MK, Ghosh S, Barnard SD, Molello JA, Reddy JK (1985) Induction of fatty acid β-oxidation and peroxisome proliferation in the liver of Rhesus monkeys by DL-040, a new hypolipidemic agent. Biochem Pharmacol 34:3473–3482

Lalwani ND, Reddy MK, Qureshi SA, Moehle CM, Hayashi H, Reddy JK (1983) Non-inhibitory effect of antioxidants ethoxyquin, 2(3)-tert-butyl-4-hydroxyanisole and 3,5,-di-tert-butyl-4-hydroxytoluene on hepatic peroxisome proliferation and peroxisomal fatty acid β-oxidation induced by a hypolipidemic agent in rats. Cancer Res 43:1680–1687

Lazarow PB, DeDuve C (1976) A fatty acyl-CoA oxidizing system in rat liver peroxisomes: enhancement by clofibrate, a hypolipidaemic drug. Proc Soc Acad Sci 73:2043–2046

Lin LI-K (1987) The use of multivariate analysis to compare peroxisome induction data on phthalate esters in rats. Toxicol Ind Health 3:25–47

Lock EA, Mitchell AM, Elcombe CR (1989) Biochemical mechanisms of induction of hepatic peroxisome proliferation. Annu Rev Pharmacol Toxicol 29:145–163

Makowska JM, Gibson GG, Bonner FW (1992) Species differences in ciprofibrate induction of hepatic cytochrome P450 4A1 and peroxisome proliferation. J Biochem Toxicol 7:183–191

Marsman DS, Popp JA (1994). Biological potential of basophilic hepatocellular foci and hepatic adenoma induced by the peroxisome proliferator, Wy-14,643. Carcinogenesis 15:111–117

Marsman DS, Cattley RC, Conway JG, Popp JA (1988) Relationship of hepatic peroxisome proliferation and replicative DNA synthesis to the hepatocarcinogenicity of the peroxisome proliferators di(2-ethylhexyl)phthalate and [4-chloro-6-(2, 3-xylidino)-2-pyrimidinylthio]acetic acid (Wy-14,643) in rats. Cancer Res 48:6739–6744

Marsman DS, Goldsworthy TL, Popp JA (1992) Contrasting hepatocytic peroxisome proliferation, lipofuscin accumulation and cell turnover for the hepatocarcinogens Wy-14,643 and clofibric acid. Carcinogenesis 13:1011–1017

Marsman DS, Swanson-Pfeiffer CL, Popp JA (1993) Lack of comitogenicity by the peroxisome proliferator hepatocarcinogens, Wy-14,643 and clofibric acid. Toxicol Appl Pharmacol 122:1–6

Moody DE, Reddy JK, Lake BG, Popp JA, Reese DH (1991) Peroxisome proliferation and nongenotoxic carcinogenesis: commentary on a symposium.
Fund Appl Toxicol 16:233–248

NCI (1976) Carcinogenesis of trichloroethylene (CAS No.79–01–6), NCI-CG-
TR-2, National Cancer Institute

NTP (1982a) Carcinogenesis bioassay of di(2-ethylhexyl)adipate (CAS
No.103–23–1) in F344 rats and B6C3F1 mice (feed study). Technical report series no 212, National Toxicology Program

NTP (1982b) Carcinogenesis bioassay of di(2-ethylhexyl)phthalate (CAS
No.117–81–7) in F344 rats and B6C3F1 mice (feed study). Technical report series no 217, National Toxicology Program

Oberhammer F, Bursch W, Parzefall W, Breit P, Erber E, Stadler M, Schulte-
Hermann R (1991) Effect of transforming growth factor β on cell death of
cultured rat hepatocytes. Cancer Res 51:2478–2485

Orton TC, Adam HK, Bentley M, Holloway B, Tucker MJ (1984) Clobuzarit:
species differences in the morphological and biochemical response of the liver
following chronic administration. Toxicol Appl Pharmacol 73:138–151

Osumi T, Hashimoto T (1978) Enhancement of fatty acyl-CoA oxidizing activity in rat liver peroxisomes by di(2-ethylhexyl)phthalate. J Biochem
83:1361–1365

Parzefall W, Erber E, Sedivy R, Schulte-Hermann R (1991) Testing for induction of DNA synthesis in human hepatocyte primary cultures by rat liver
tumor promoters. Cancer Res 51:1143–1147

Popp JA, Cattley RC (1993) Peroxisome proliferators as initiators and promoters of rodent hepatocarcinogenesis. In: Gibson G, Lake B (eds) Peroxisomes: biology and importance in toxicology and medicine. Taylor and
Francis, London, pp 653–665

Popp JA, Marsman DS (1991) Chemically-induced cell proliferation in liver
carcinogenesis. In: Butterworth BE, Slaga TJ, Farland W, McClain M (eds)
Chemically induced cell proliferation: implications for risk assessment,
Wiley-Liss, New York, pp 389–395

Price RJ, Evans JG, Lake BG (1992) Comparison of the effects of nafenopin
on hepatic peroxisome proliferation and replicative DNA synthesis in the
rat and Syrian hamster. Food Chem Toxicol 30:937–944

Randerath E, Randerath K, Reddy R, Danna TF, Rao MS, Reddy JK (1991)
Induction of rat liver DNA alterations by chronic administration of peroxisome proliferators as detected by 32P-postlabelling. Mut Res 247:65–76

Rao MS, Lalwani ND, Watanabe TK, Reddy JK (1984) Inhibitory effects of
antioxidants ethoxyquin and 2(3)-tert-butyl-4-hydroxyanisole on hepatic
tumorigenesis in rats fed ciprofibrate, a peroxisome proliferator. Cancer
Res 44:1072–1076

Reddy JK, Lalwani ND (1983) Carcinogenesis by hepatic peroxisome proliferators: evaluation of the risk of hypolipidemic drugs and industrial plasticizers to humans. CRC Crit Rev Toxicol 12:1–58

Reddy JK, Rao MS (1989) Oxidative DNA damage caused by persistent peroxisome proliferation: its role in hepatocarcinogenesis. Mutat Res 214:63–68

Reddy JK, Lalwani ND, Qureshi SA, Reddy MK, Moehle CM (1984) Induction of hepatic peroxisome proliferation in non-rodent species, including primates. Am J Pathol 114:171–183

Reddy JK, Reddy MK, Usman MI, Lalwani ND, Rao MS (1986) Comparison of hepatic peroxisome proliferative effect and its implication for hepatocarcinogenicity of phthalate esters, di(2-ethylhexyl)phthalate and di(2-ethylhexyl)adipate with a hypolipidemic drug. Environ Health Perspect 65:317–327

Rhodes C, Orton TC, Pratt IS, Batten PL, Bratt H, Jackson SJ, Elcombe CR (1986) Comparative pharmacokinetics and subacute toxicity of di(2-ethylhexyl)phthalate (DEHP) in rats and marmosets: extrapolation of effects in rodents to man. Environ Health Perspect 65:299–308

Rodricks JV, Turnbull D (1987) Interspecies differences in peroxisomes and peroxisome proliferation. Toxicol Ind Health 3:197–212

Rumsby PC, Davies MJ, Price RJ, Lake BG (1994) Effect of some peroxisome proliferators on transforming growth factor-β1 gene expression and insulin-like growth factor II/mannose-6-phosphate receptor gene expression in rat liver. Carcinogenesis 15:418–421

Sakuma M, Yamada J, Suga T (1992) Comparison of the inducing effect of dehydroepiandrosterone on hepatic peroxisome proliferation-associated enzymes in several rodent species. A short-term administration study. Biochem Pharmacol 43:1269–1273

Sharma R, Lake BG, Foster J, Gibson GG (1988a) Microsomal cytochrome P-452 induction and peroxisome proliferation by hypolipidaemic agents in rat liver. A mechanistic inter-relationship. Biochem Pharmacol 37:1193–1201

Sharma R, Lake BG, Gibson GG (1988b) Co-induction of microsomal cytochrome P-452 and the peroxisomal fatty acid β-oxidation pathway in the rat by clofibrate and di-(2-ethylhexyl)phthalate. Dose-response studies. Biochem Pharmacol 37:1203–1206

Short RD, Robinson EC, Lington AW, Chin AE (1987) Metabolic and peroxisome proliferation studies with di(2-ethylhexyl)phthalate in rats and monkeys. Toxicol Ind Health 3:185–194

Stott WT (1988) Chemically induced proliferation of peroxisomes: implications for risk assessment. Regul Toxicol Pharmacol 8:125–159

Takagi A, Sai K, Umemura T, Hasegawa, R, Kurokawa Y (1990a) Significant increase of 8-hydroxydeoyguanosine in liver DNA of rats following short-term exposure to the peroxisome proliferators di(2-ethylhexyl)phthalate and di(2-ethylhexyl)adipate. Jpn J Cancer Res 81:213–215

Takagi A, Sai, K, Umemura T, Hasegawa, R, Kurokawa Y (1990b) Relationship between hepatic peroxisome proliferation and 8-hydroxydeoyguanosine formation in liver DNA of rats following long-term exposure to three peroxisome proliferators; di(2-ethylhexyl)phthalate, aluminium clofibrate and simfibrate. Cancer Lett 53:33–38

Tamura H, Iida T, Watanabe T, Suga T (1991) Lack of induction of hepatic DNA damage on long-term administration of peroxisome proliferators in male F-344 rats. Toxicology 69:55–62

Thornber CW (1979) Isosterism and molecular modification in drug design. Chem Soc Rev 8:563–580

Tucker MJ, Orton TC (1993) Toxicological studies in primates with three fibrates. In: Gibson G, Lake B (eds) Peroxisomes: biology and importance in toxicology and medicine. Taylor and Francis, London, pp 425–447

Von Däniken A, Lutz WK, Schlatter C (1981) Lack of covalent binding to rat liver DNA of the hypolipidemic drugs clofibrate and fenofibrate. Toxicol Lett 7:311–319

Von Däniken A, Lutz WK, Jäckh R, Schlatter C (1984) Investigation of the potential for binding of di(2-ethylhexyl)phthalate (DEHP) and di(2-ethylhexyl)adipate (DEHA) to liver DNA in vivo. Toxicol Appl Pharmacol 73:373–387

Wada N, Marsman DS, Popp JA (1992) Dose-related effects of the hepatocarcinogen, Wy-14,643 on peroxisomes and cell replication. Fund Appl Toxicol 18:149–154

Watanabe T, Horie S, Yamada J, Isaji M, Nishigaki T, Naito J, Suga T (1989) Species differences in the effects of bezafibrate, a hypolipidemic agent, on hepatic peroxisome-associated enzymes. Biochem Pharmacol 38:367–371

Yeldandi AV, Milano M, Subbarao V, Reddy JK, Rao MS (1989) Evaluation of liver cell proliferation during ciprofibrate-induced hepatocarcinogenesis. Cancer Lett 47:21–27

11 Peroxisome Proliferators Mimic an Endogenous Inducer and Inactivate a Transcriptional Repressor in Bacillus megaterium

N. English, V. Hughes and C. R. Wolf

11.1 Introduction

11.1.1 *Bacillus megaterium* – A Prokaryotic Model
for Cytochrome P450 Induction

Cytochrome P450s are a large family of monooxygenase enzymes involved in the metabolism of a wide variety of xenobiotics and natural
compounds (Coney 1967). These compounds, which include the barbiturates, aromatic hydrocarbons and peroxisome proliferators, also induce cytochrome P450s. The prokaryote, *Bacillus megaterium* ATCC
14581, possesses a P450 (BM-3) active in fatty acid metabolism which
introduces hydroxyl groups at the ω-1,2,3 position of fatty acids. BM-3
is only expressed in late log/early stationary cultures but, intriguingly, is
induced by barbiturates and their corresponding analogues in early log
cultures (Kim and Fulco 1983). We decided to study the megaterium
system with a view to dissecting out the factors that are important in
initiating the induction event and to determine whether any common
denominators exist between barbiturates and peroxisome proliferators.
This is particularly pertinent in light of the fact that BM-3 has a high
level of identity to the mammalian CYP4A family whose members are
strongly induced by peroxisome proliferators. In fact, the CYP 4A
family has a higher level of identity to BM-3 than to other mammalian
cytochrome families, which suggests they arose from a common ances-
tral lineage (Nebert et al. 1989). Thus, common elements of regulation
may exist that have also been conserved throughout evolution and may
be dissected in the genetically amenable model of *Bacillus megaterium*.

11.1.2 Mechanism of BM-3 Induction
Involves a Transcriptional Repressor Bm3R1

The mechanism of barbiturate-mediated P450 BM-3 induction has been
partially characterised and the promoter region of the BM-3 operon has
been dissected (Shaw and Fulco 1992). This work identified a repressor
protein termed Bm3R1 involved in the regulation of this gene (Shaw
and Fulco 1993). Figure 1 depicts the basic control elements which have
been proposed for this operon. A palindromic 20-bp operator site is
located just upstream of the open reading frame of Bm3R1 and interacts

1. Transcriptional Repression in the Absence of Inducers

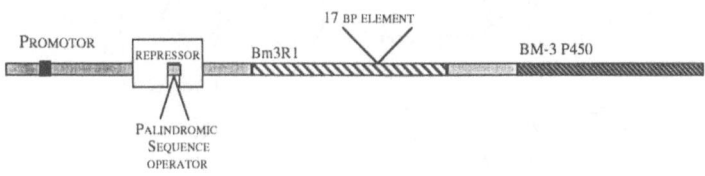

2. De-Repression of Transcription by Addition of Barbiturates or Peroxisome Proliferators

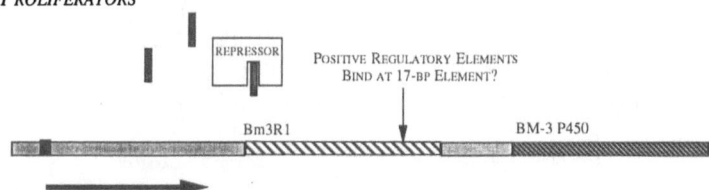

Fig. 1. Model of transcriptional activation of the BM-3 operon. Schematic diagram of barbiturate and peroxisome proliferator-mediated induction of BM-3 in *Bacillus megaterium*. In the absence of inducer, the repressor (Bm3R1) binds to the operator, a 20-bp palindromic sequence, and prevents initiation of transcription at the promoter of the gene. Barbiturates or peroxisome proliferators interact with Bm3R1 and prevent its binding to the operator. Transcription is then initiated with the formation of a bi-cistronic message encoding Bm3R1 and P450 BM-3

with the Bm3R1 protein. It is thought that under noninducing conditions Bm3R1 binds tightly to this sequence to sterically inhibit transcriptional initiation at the promoter, which is situated 11 bp downstream of the palindromic operator. It is believed that barbiturates interact with Bm3R1 and abrogate its binding to this site, which may in turn allow binding of other positive factors to initiate transcriptional activation of this operon. The importance of Bm3R1 was clearly demonstrated by the existence of a mutant which harboured a single point mutation in the region of the gene encoding the helix-loop-helix DNA binding domain of the protein. Mutants expressed P450 BM-3 levels in excess of that

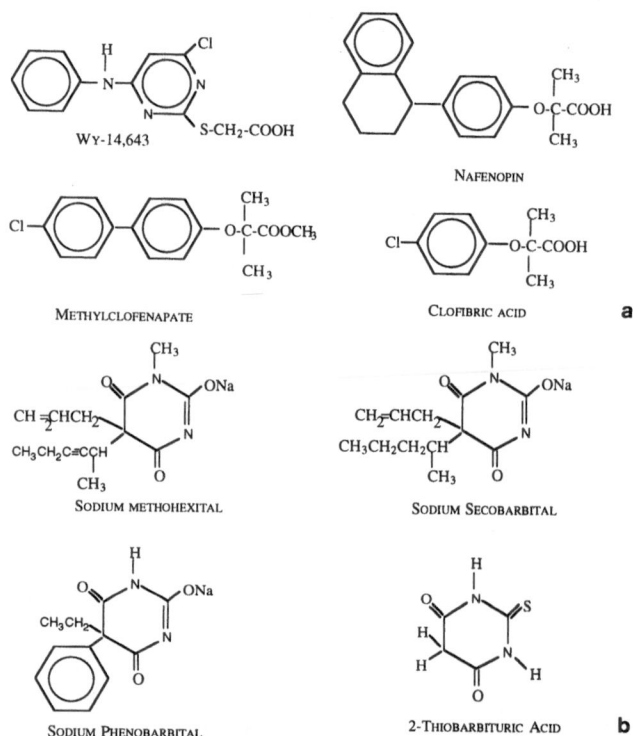

Fig. 2a,b. Chemical structures of the xenobiotics. Structures of the peroxisome proliferators (**a**) used in these studies and some of the more powerful barbiturate inducers (**b**) are presented. 2-Thiobarbituric acid, a barbiturate which does not induce BM-3, is also shown

obtained by treatment of wild-type cells with the barbiturate inducer pentobarbital (Wen and Fulco 1987).

11.1.3 Peroxisome Proliferators Are Novel Inducers of BM-3

We tested whether the peroxisome proliferators could also induce BM-3 in *Bacillus megaterium* and have found that a variety of peroxisome proliferators are strong inducers, being more potent than the barbiturates in all cases. Furthermore, we provide evidence that these compounds interact directly with Bm3R1 and dissociate it from its DNA binding site. The ability of these compounds to induce P450 expression is directly related to their relative potency to dissociate the Bm3R1–DNA interaction. The structures of some peroxisome proliferators and the more potent barbiturate inducers along with a noninducing barbiturate, 2-thiobarbituric acid, are shown in Fig. 2.

11.1.4 Insights into the Nature of the Endogenous Inducer of BM-3

Fatty acids play multiple roles in cells; they are integral to the plasma membrane, and in higher organisms are the precursors of important biologically active compounds such as prostanoids and also become covalently linked to proteins in post-translational modifications. More recently, there have been several reports of fatty acids modulating gene expression (McDonough et al. 1992; Distel et al. 1992); however, there has only been one report of fatty acyl CoA esters directly interacting with a transcription factor, FadR (DiRusso et al. 1992), which acts as both a positive and negative regulator of gene expression in *Escherichia coli* (Henry and Cronan 1991). Since peroxisome proliferators are known to be closely associated with fatty acid metabolism in mammalian systems (Elcombe and Mitchell 1986; Dreyer et al. 1993), we postulate that the endogenous inducers of this bacterial P450 are fatty acids and provide evidence that unsaturated fatty acids can dissociate the Bm3R1–DNA complex at physiological concentrations in vitro.

    This work provides the first evidence that peroxisome proliferators interact directly with a prokaryotic transcription factor to mediate their effects on gene expression in *Bacillus megaterium* and suggests that they mimic an increase in intracellular fatty acid levels, which normally occurs during stationary phase – presumably a time when fatty acids become mobilised.

11.2 Materials and Methods

11.2.1 Growth and Harvesting of Cells

Bacillus megaterium cells were grown at 35°C with aeration either in the presence or absence of drug, to an optical density of 0.6, at 600 nm in glucose salts medium (Grelet 1951). Cultures were grown for approximately 8 h. Cells were harvested by centrifugation at 2000 g, washed in 0.1 M phosphate buffer pH 7.4, centrifuged again, then suspended in the same buffer and lysed by sonication. 30 000 g supernatants were used for immunoblot analysis and enzyme assay, 100 000 g supernatants were used for spectroscopic P450 quantitation as described previously (Omura and Sato 1964).

11.2.2 Enzyme Assays

Fatty acid hydroxylase activities associated with BM-3 were monitored by measuring the rate of nicotinamide adenine dinucleotide phosphate, reduced (NADPH) oxidation using sodium palmitate as the cosubstrate, as previously described (Ruettinger and Fulco 1981).

11.2.3 Immunoblots

Cell extracts for immunoblot analysis (Towbin et al. 1979) were separated by sodium dodecyl sulphate-polyacrylamide gel electrophoresis (SDS-PAGE; Laemmli 1970) and transferred to nitrocellulose. BM-3 was visualised using a polyclonal antisera raised in rabbits to the reductase domain of this protein.

11.2.4 Electrophoretic Mobility Shift Assay

Electrophoretic mobility shift assays (EMSAs) were carried out according to established protocols (Ausubel et al. 1993). A double-stranded oligonucleotide encompassing the binding site of Bm3R1 (5'-CGGAATGAACGTTCATTCCG-3') was incubated with 4.8 μg of recombinant Bm3R1, either in the presence or absence of test drug. Samples were then electrophoresed on 4% gel, dried and exposed to X-ray film.

11.2.5 Two-Dimensional Protein Analysis
of *Bacillus megaterium* Proteins

Bacillus megaterium were grown either in the presence or absence of 100 μM nafenopin as described above. Sonicated 30 000 g cell extracts were obtained and two-dimensional electrophoresis was performed essentially as described previously (O'Farrel 1975). Ampholines with a pH range of 5–7 and 3.5–10 in a ratio of 4:1 were used in isoelectric focusing in the first dimension and a 12.5% acrylamide SDS-PAGE in the second dimension (Laemmli 1970).

11.3 Results

Figure 3a displays the P450 spectrum of cytosolic extracts of cells treated with the potent peroxisome proliferator nafenopin and a solvent-treated control. The nafenopin treated extract had P450 levels sevenfold higher than control. A protein of molecular mass 120 kDa was detected on immunoblots shown in Fig. 3b that corresponded to the expected size of BM-3. A variety of other peroxisome proliferators were also tested by monitoring fatty acid oxidation rates. Figure 4 shows the relative potency of a number of other peroxisome proliferators, including Wy,14643 and methylclofenapate, compared to secobarbital – one of the more powerful barbiturate inducers. In Fig. 4 it is clearly demonstrated that peroxisome proliferators are much more potent inducers of BM-3 than the barbiturates, Wy,14643, nafenopin and methylclofenapate being 48-, 22- and 2.7-fold more effective than secobarbital on a molar basis.

We speculated that peroxisome proliferators may be acting in an analogous manner to barbiturates by interacting with Bm3R1 to activate BM-3 transcription. In order to test whether this was the case, His-tagged protein was cloned, expressed in *E. coli* and purified to apparent homogeneity on a nickel–agarose column. This protein appeared as a

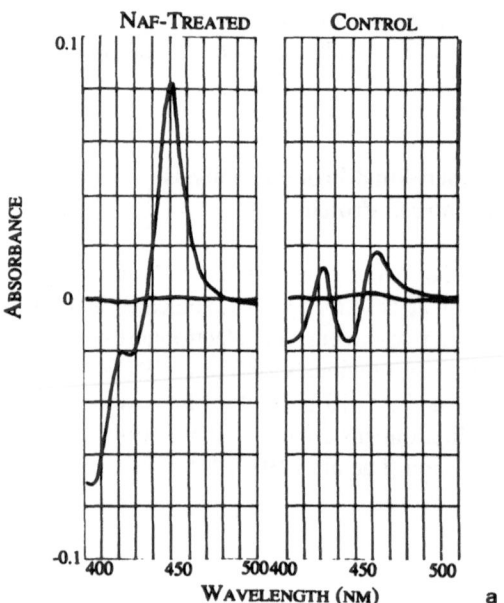

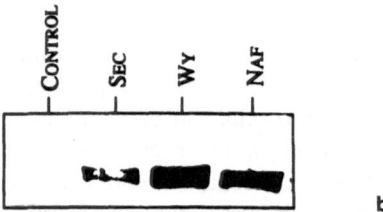

Fig. 3a,b. Induction of BM-3 by peroxisome proliferators. **a** Cells were grown in the presence of nafenopin (*NAF*, 100 μ*M*) or the solvent dimethylsulphoxide (0.24%)(*control*). Cytochrome P450 content was calculated from the absorbance at 450 nm. **b** Immnunoblots of soluble extracts of *Bacillus megaterium* either untreated (*Control*) or treated with: 2 m*M* secobarbital (*SEC*), 50 μ*M* Wy14643 (*WY*) and 200 μ*M* nafenopin (*NAF*)

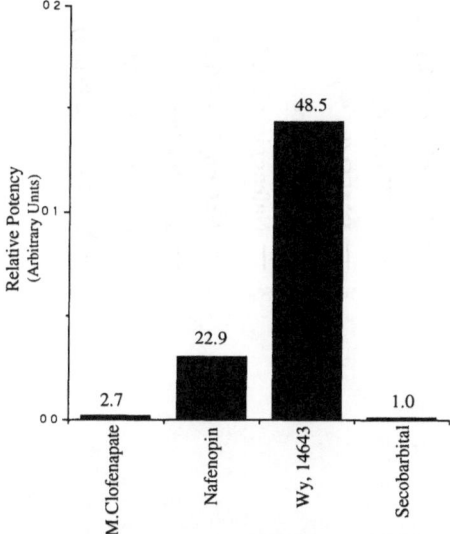

Fig. 4. Relative potency of drugs to induce BM-3. Relative potency of three peroxisome proliferators – methylclofenapate, nafenopin and Wy,14643 – and secobarbital, one of the more powerful barbiturates. Potency is presented as the inverse of the concentration of test compound required to give half maximal induction (as determined by enzymic assay)

single band on an SDS gel (data not shown) and had an apparent molecular mass of 23 kDa, in close agreement with the calculated molecular weight predicted from its amino acid sequence. Recombinant protein was used in gel retardation assays together with an oligonucleotide encoding the putative binding site for Bm3R1. Figure 5 clearly shows that the recombinant protein formed a specific complex with the oligonucleotide probe. Interestingly, the peroxisome proliferators nafenopin and Wy,14643 inhibited the formation of this complex in a dose-dependent manner, and it was shown in further experiments (manuscript in preparation) that they were at least an order of magnitude more effective than secobarbital. However, 2-thiobarbituric acid, deemed to be a noninducer of BM-3, was incapable of dissociating the Bm3R1/oligonucleotide operator complex. This agreed well with the previously observed in vivo potencies of these agents to induce BM-3

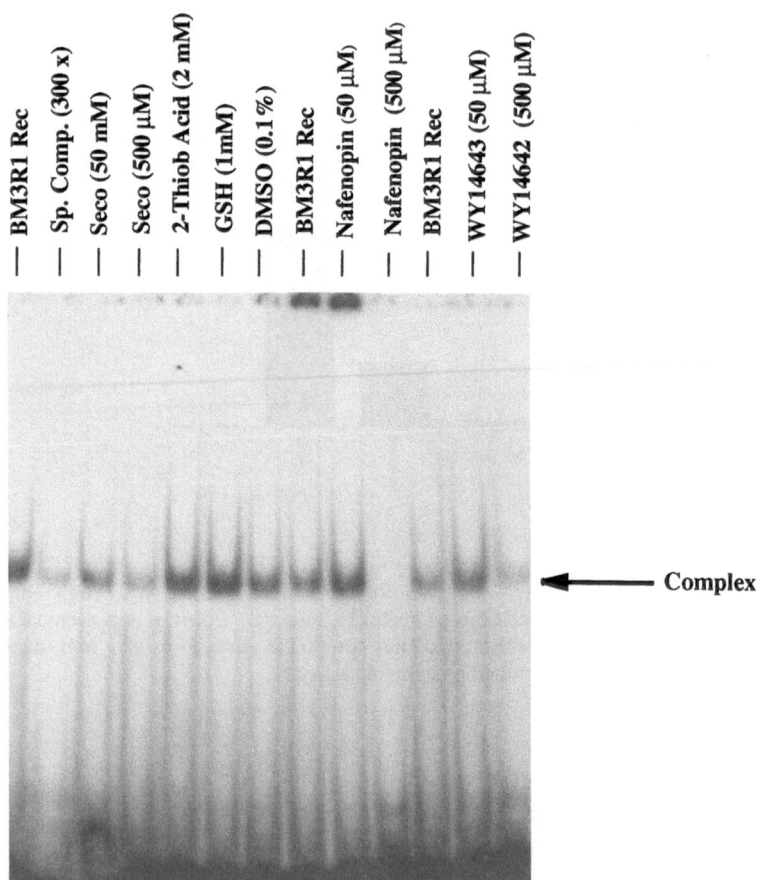

Fig. 5. Binding of Bm3R1 to the palindromic sequence: inhibition with barbiturates and peroxisome proliferators. Electrophoretic mobility shift assays demonstrate the binding of the 20 bp palindromic sequence to recombinant Bm3R1. Assays were also carried out in the presence of a 300-fold excess of specific competitor (*Sp. Comp.*) (i.e. the unlabelled palindromic operator sequence). In addition, a number of compounds were included in the incubation mix: *Seco,* secobarbital; *2-Thiob acid*, 2-thiobarbituric acid; *GSH*, reduced glutathione; *DMSO,* dimethylsulphoxide

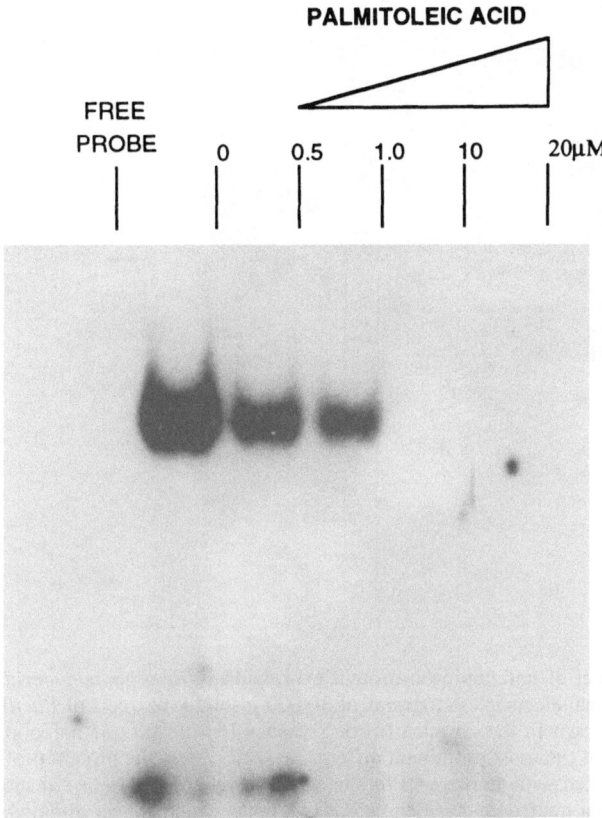

Fig. 6. Inhibition of binding of Bm3R1 to the palindromic sequence by palmitoleic acid. Electrophoretic mobility shift assays of the 20-bp palindromic sequence in the absence or presence of increasing concentrations of palmitoleic acid

expression. This indicated that peroxisome proliferators interact with the repressor to dissociate it from its operator complex. It will be interesting to determine how these chemically diverse compounds interact with Bm3R1 when no obvious structural motif is apparent.

Since P450 BM-3 is involved in the hydroxylation of fatty acids and is induced by a range of peroxisome proliferators which are also associated with lipid metabolism, we examined the ability of a variety of fatty

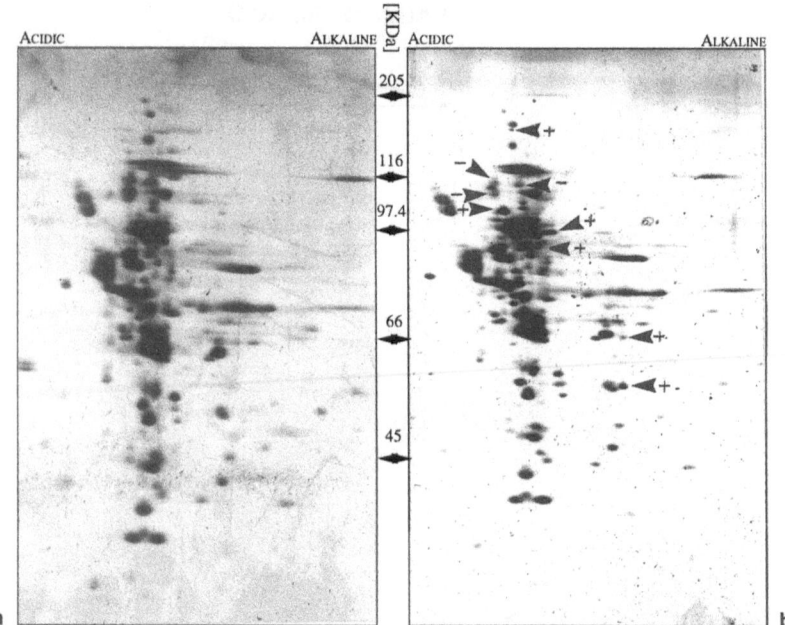

Fig. 7a,b. Effect of nafenopin on protein expression in *Bacillus megaterium*. Two-dimensional electrophoretograms of 30 000 *g* soluble extracts of *Bacillus megaterium* grown in the absence (**a**) or presence (**b**) of 100 µ*M* nafenopin. The positions of the most prominent differences observed in the protein profile have been marked with an *arrow*. Proteins whose level appears to be enhanced in the nafenopin-treated cells is denoted as (+), those which appear diminished as (-)

acids to dissociate the repressor protein from its operator sequence. Figure 6 shows that palmitoleate has the ability to dissociate this complex at concentrations between 1 and 10 µ *M*. Other unsaturated fatty acids such as arachidonate could also dissociate the complex at concentrations in the nanomolar range (data not shown). These observations strongly suggest that fatty acids are the physiological inducers of this P450.

Figure 7 shows a two-dimensional analysis of the protein profile of *Bacillus megaterium* which was grown either in the presence or absence of nafenopin. A number of proteins appear to be present at higher levels in the nafenopin-treated sample, suggesting that other protein species of unknown identity may also be induced. Intriguingly, there are also a number of proteins in the profile that are under-represented. Thus it appears that nafenopin is acting in both a positive and negative manner to affect the expression of proteins. It remains to be determined whether all these effects are due to nafenopin modulating the transcriptional control of Bm3R1.

11.4 Discussion

11.4.1 Peroxisome Proliferators and Barbiturates Have Common Overlapping Effects

We have shown that peroxisome proliferators are novel and potent inducers of P450 BM-3 in *Bacillus megaterium*. Futhermore, we have demonstrated that barbiturates and peroxisome proliferators interact with the same transcription factor to induce BM-3 expression and provide the first strong evidence that peroxisome proliferators act directly on Bm3R1 to induce P450 expression in this bacterial system. This has important implications because barbiturates and peroxisome proliferators, previously thought to be distinct classes of inducing agents in mammalian systems, may, in fact, have common overlapping metabolic effects. It remains to be determined whether barbiturates have a similar effect on the protein profile of *Bacillus megaterium* and whether the positive and negative effects on expression of protein are due solely to Bm3R1. If this is the case, then it appears that Bm3R1 may be acting in a similar manner to FadR (Henry and Cronan 1992) and thus be the second example of prokaryotic transcription repressed by ligand inactivation of a positive factor.

11.4.2 Are Fatty Acids the Endogenous Ligands of Bm3R1?

We have shown that unsaturated fatty acid molecules can dissociate the Bm3R1–operator complex at physiological levels and therefore have the potential to act as the endogenous inducers of BM-3. There are an increasing number of reports in the literature which demonstrate that unsaturated fatty acids have an ability to activate the mammalian perox-isome proliferator activated receptor (PPAR; Keller et al. 1993), a member of the steroid hormone receptor superfamily. This receptor is also activated by peroxisome proliferators (Issemann and Green 1990). Thus, there are strong grounds for postulating that this mammalian receptor has evolved to monitor intracellular fatty acid levels of the cell. The fact that in *Bacillus megaterium* P450 BM-3 functions as a fatty acid hydroxylase and is dramatically induced by peroxisome prolif-erators, as well as the observation that unsaturated fatty acids can dissociate Bm3R1 from its operator site at such low concentrations, suggests that Bm3R1 may have evolved to fulfil a similar function in *Bacillus megaterium*. We believe this is the first documented example of free fatty acids directly interacting with a transcription factor to modulate gene expression.

11.4.3 Evolutionary Conservation of the Peroxisome Proliferator Response

There seem to be a number of apparent similarities between the way these compounds exert their effects in *Bacillus megaterium* and mam-malian systems, which suggests that features of this response may have been conserved throughout evolution. However, as there is no sequence homology between PPAR and Bm3R1 it seems likely that these proteins have evolved separately but have converged to have similar functional roles in sensing changes in fatty acid levels within cells. This hypothesis is supported by the fact that PPAR is activated by both fatty acids and peroxisome proliferators, although no direct binding of these molecules to PPAR has been demonstrated to date (Green 1992).

11.4.4 Are There Common Links Between the Mode of Action of Barbiturates and Peroxisome Proliferators?

This work also has implications for understanding P450 induction by barbiturates. The molecular basis for barbiturate-mediated induction in higher organisms is very poorly understood. It seems apparent that there are common overlapping features in the induction of cytochromes by barbiturates and peroxisome proliferators. For example, the classical peroxisome proliferator clofibric acid has also been documented to induce CYP2B1 in murine liver (Bars et al. 1989, 1993), an enzyme regulated by barbiturate-like inducers, which indicates that there is a relationship between these two classes of xenobiotic in their ability to regulate mammalian P450 genes.

How then can we rationalise the ability of structurally diverse molecules to interact with the same transcription factor? Close examination of the barbiturate structures reveals that there is a strong positive correlation between inducibility and the presence of an aliphatic side chain on the barbiturate ring (see Fig. 2). For example, a barbiturate analogue 2-thiobarbituric acid lacking this side chain is inactive in the megaterium system; however, secobarbital which possesses a straight aliphatic side chain has a potency of 33.3 relative to phenobarbital, which has an aromatic side chain. Furthermore, methohexital, similar in structure to secobarbital but with an unsaturated side chain, has a relative potency of 71 (Kim and Fulco 1983). The inducing peroxisome proliferators also possess an aliphatic group in their structures. Thus we postulate that a possible common denominator linking these inducing agents is their ability to perturb or mimic a fatty acid-like species, which in turn can modulate the expression of these genes. We are currently testing whether this hypothesis is tenable.

References

Ausubel FM, Brent R, Kingston RE, Moore DD, Seidman JG, Smith JA, Struhl K (1993) Current protocols in molecular biology, vol 1, section 12. Wiley Interscience, New York

Bars RG, Mitchell AM, Wolf CR, Elcombe CR (1989) Induction of cytochrome P-450 in cultured rat hepatocytes. The heterogeneous localisation of specific isoenzymes using immunocytochemistry. Biochem J 262:151–158

Bars RG, Bell DR, Elcombe CR (1993) Induction of cytochrome P-450 and peroxisomal enzymes by clofibric acid in vivo and in vitro. Biochem Pharmacol 45:2045–53

Conney AH (1967) Pharmacological implications of microsomal enzyme induction. Pharmacol Rev 19:317–366

DiRusso CC, Heimart TL, Metzger AK (1992) Characterisation of FadR, a global transcriptional regulator of fatty acid metabolism in E. coli. J Biol Chem 267:8685–8691

Distel RJ, Robinson GS, Spiegelman BM (1992) Fatty acid regulation of gene expression. J Biol Chem 267:5937–5941

Dreyer C, Keller H, Mahfoudi A, Laudet V, Krey G, Wahli W (1993) The control of peroxisomal β-oxidation pathway by a novel family of nuclear hormone receptors. Biol Cell 77:67-76

Elcombe CR, Mitchell AM (1986) Peroxisome proliferation due to di(2-ethyl hexyl phthalate (DEHP): species differences and possible mechanisms. Environ Health Perspect 70:211–219

Green S (1992) Receptor-mediated mechanisms of peroxisome proliferators. Biochem Pharmacol 43:393–401

Grelet N (1951) La determinisme de la sporulation de Bacillus megaterium. Ann Inst Pasteur 81:430–40

Henry MF, Cronan JE (1991) Escherichia coli transcription factor that both activates fatty acid synthesis and represses fatty acid degradation. J Mol Biol 222:843–849

Henry MF, Cronan J.E. (1992) A new mechanism of transcriptional regulation: release of an activator triggered by small molecule binding. Cell 70:671–679

Issemann I , Green S (1990) Activation of a member of the steroid hormone receptor superfamily by peroxisome proliferators. Nature 347:211–219

Keller B, Dreyer C, Medin J, Mahfoudi A, Ozata K, Wahli W (1993) Fatty acids and retinoids control lipid metabolism through activation of peroxisome proliferator-activated receptor-retinoid X receptor heterodimers. Proc Natl Acad Sci U S A 90:2160–2164

Kim B, Fulco AJ (1983) Induction by barbiturates of a cytochrome P-450-dependent fatty acid monooxygenase in Bacillus megaterium: relationship between barbiturate structure and inducer activity. Biochem Biophys Res Commun 116:843–850

Laemmli UK (1970) Cleavage of structural proteins during the assembly of the head of bacteriophage T4. Nature 227:680–685

McDonough VM, Stukey JE, Martin CE (1992) Specificity of unsaturated fatty acid-regulated expression of the Saccharaomyces cerevisiae OLE1 gene. J Biol Chem 267:5931–5936

Nebert DW, Nelson DR, Feyereisen R (1989) Evolution of the cytochrome P-450 genes. Xenobiotica 19:1149–1160

O'Farrel PH (1975) High resolution two-dimensional electrophoresis of proteins. J Biol Chem 250:4007–4021

Omura T, Sato RJ (1964) The carbon monoxide-binding pigment of liver microsomes. Evidence for its haemoprotein nature. J Biol Chem 239:2370–2378

Ruettinger RT, Fulco AJ (1981) Epoxidation of unsaturated fatty acids by a soluble cytochrome P450-dependent system from Bacillus megaterium. J Biol Chem 256:5728–5734

Shaw G, Fulco AJ (1992) Barbiturate-mediated regulation of expression of the cytochrome P450 BM-3 gene of Bacillus megaterium by Bm3R1 protein. J Biol Chem 267:5515–5526

Shaw GC, Fulco AJ (1993) Inhibition by barbiturates of the binding of BM3R1 repressor to its operator site on the barbiturate-inducible cytochrome P450 BM3 gene of Bacillus megaterium. J Biol Chem 268:2997–3004

Towbin H, Staehelin T, Gordon J (1979) Electrophoretic transfer of proteins from polyacrylamide gels to nitrocellulose sheets. Proc Natl Acad Sci U S A 76:4350-4353

Wen LP , Fulco AJ (1987) Cloning of the gene encoding a catalytically self-sufficient cytochrome P-450 fatty acid monooxygenase induced by barbiturates in Bacillus megaterium and its functional expression and regulation in heterologous (Escherichia coli) and homologous (Bacillus megaterium) hosts. J Biol Chem 262:6676–6682

12 The Interaction of Genes and Hormones in Murine Hepatocarcinogenesis

N. R. Drinkwater

12.1 Introduction

Considerable progress has been made in the molecular identification of human "cancer genes," such as those resulting in the heritable development of retinoblastoma, Wilm's tumor, or familial adenomatous polyposis (Friend et al. 1986; Rose et al. 1990; Kinzler et al. 1991). However, human genetic studies have been less effective in identifying "risk-modifier" genes that may increase by 5- to 50-fold an individual's susceptibility to the development of more common malignancies because of the low effective penetrance of such genes and the large variation in environmental factors that influence cancer development in humans. Animal models provide a powerful approach to the identification and characterization of such risk-modifier genes. There is substantial variation among inbred strains of mice and rats in their susceptibilities to spontaneous and chemically induced carcinogenesis at a variety of tissue sites (Drinkwater and Bennett 1991), and the genetic basis for this variation in risk can be studied under well-defined ex-

Table 1. Strain variation in susceptibility to hepatocarcinogenesis in mice

Strain	ENU[a]			DEN[b]		
	Number of mice	Tumor multiplicity[c]		Number of mice	Tumor multiplicity[c]	
SWR/J	16	0		24	0.8	(1.3)
A/J				22	1.5	(1.9)
C57BL/6J	25	0.3	(0.7)	23	1.4	(1.6)
AKR/J				18	7	(5)
BALB/cByJ				23	9	(6)
C57BR/cdJ	23	7.1	(7.0)	32	37	(23)
P/J	16	15	(9)			
SM/J	15	17	(14)			
DBA/2J				23	55	(20)
CBA/J	24	45	(14)			
C3H/HeJ				22	78	(30)

[a]Data from Kemp and Drinkwater (1989). Male mice were given a single injection of ENU (0.25 μmol/g body weight) at 12 days of age and sacrificed at 32 weeks of age.
[b]Data from unpublished experiments by M. Bennett, T. Poole, M. Winkler and N. Drinkwater. Male mice were treated at 12 days of age with diethylnitrosamine (0.1 μmol/g body weight) and sacrificed at 32 weeks of age.
[c]Values in the table are mean tumor multiplicity (standard deviation).

perimental conditions. Understanding the mechanism of action of specific genes that control the risk of inbred mice for tumor induction will provide insights into the underlying biology of carcinogenesis and paradigms for understanding the action of risk-modifier genes in humans.

Genetic factors play an important role in determining the susceptibility of inbred mice to both spontaneous and chemically induced hepatocarcinogenesis (Drinkwater 1989; Drinkwater and Bennett 1991). We have compared directly ten inbred mouse strains for their sensitivities to hepatocarcinogenesis by treatment of preweanling male mice with a single injection of N-ethyl-N-nitrosourea (ENU) (Kemp and Drinkwater 1989) or diethylnitrosamine (DEN; Table 1) (M. Bennett, T. Poole, M. Winkler, and N. Drinkwater, unpublished results). Male C3H/HeJ, CBA/J, and DBA/2J mice were the most susceptible with a mean tumor multiplicity in DBA/2J mice that was approximately 70% that for the two former strains. A moderate tumor yield (10%–30% that for C3H mice) was obtained for AKR/J, BALB/cByJ, SM/J, and C57BR/cdJ mice, while A/J, C57BL/6J, and SWR/J mice were highly resistant to hepatocarcinogenesis. With the exception of the DBA/2 mouse, the relative susceptibilities of the above strains to ENU- or DEN-induced liver tumors were similar to those reported for other carcinogen treatment protocols and for the spontaneous incidence of liver tumors in male mice. Although DBA/2 mice are nearly as susceptible as C3H mice when treated perinatally with carcinogen, this strain has a low spontaneous incidence of liver tumors (Drinkwater 1989) and is relatively resistant to the induction of liver tumors by carcinogen treatment of adult animals (Diwan et al. 1986).

Hepatocarcinogenesis in mice is also under strong hormonal control. It has long been known that male mice are significantly more susceptible than females to spontaneous liver tumors and to liver tumor induction by perinatal treatment with carcinogens (Andervont 1950; Vesselinovitch et al. 1980). This sexual dimorphism results from the contrasting effects of promotion of hepatocarcinogenesis by androgens and inhibition of liver tumor induction by the female hormonal environment. Early studies by Vesselinovitch and coworkers (Vesselinovitch et al. 1980, 1982) demonstrated that castration of male mice resulted in a decrease in the yield of liver tumors while ovariectomy of females increased the occurrence of hepatic neoplasms. Analysis of the development of preneoplastic lesions revealed that these hormonal effects on tumor development resulted from the influences of the sex hormones on the growth of these precursors to hepatic neoplasms (Vesselinovitch et al. 1982; Moore et al. 1981; Hanigan et al. 1988).

We have pursued a genetic approach to understanding the mechanisms by which the hormonal environment of the host controls the development of liver tumors in mice. The availability of mutant mice that are defective in the expression of the androgen receptor allowed us to study the role of this receptor in the regulation of preneoplastic growth by androgens. The identification of an inbred mouse strain that is insensitive to the inhibitory effects of the female hormonal environment has provided us with new tools with which to understand the genetic basis of this pathway of growth regulation.

12.2 The Role of the Androgen Receptor
in Hepatic Tumor Promotion

As noted above, comparison of intact and castrated male mice for the development of preneoplastic and neoplastic hepatic lesions indicated strongly that the male hormonal milieu is a promoter of hepatocarcinogenesis (Vesselinovitch et al. 1982). In ovariectomized female B6C3F1 mice, chronic treatment with testosterone increased significantly the yield of liver tumors (Moore et al. 1981; Kemp et al. 1989). In order to determine whether the promoting effects of this steroid hormone depended on binding to the androgen receptor, we studied liver tumor development in *Tfm* (Testicular feminization) mutant mice.

The X-linked *Tfm* locus encodes the androgen receptor; mutant (*Tfm*/Y) male mice fail to express androgen receptor mRNA or functional protein and are defective for a variety of male-specific phenotypes (Lubahn et al. 1988; Young et al. 1989). These mutant male mice (*Tfm*/Y) are much less susceptible to liver tumor induction than wild-type animals (Kemp et al. 1989). Treatment of wild-type, male C57BL/6J with DEN (0.2 μmol/g body weight) at 12 days of age resulted in a mean liver tumor multiplicity of 20 ± 10 at 50 weeks of age; the yield of liver tumors in similarly treated *Tfm* mutant mice was 0.7 ± 1.4. Chronic treatment of the mutant mice with supraphysiological doses of testosterone did not result in a substantial increase in liver tumor induction. The rate of growth of glucose-6-phosphatase-deficient hepatic foci in *Tfm*/Y mice was approximately one half that observed in wild-type males, indicating that the receptor plays a role in controlling the proliferation of preneoplastic hepatocytes. Together with previous studies of the effect of castration on hepatocarcinogenesis, these experiments demonstrate that a functional androgen receptor is required for the promotion of hepatocarcinogenesis by testosterone.

The androgen receptor is expressed at significant levels in the livers of normal mice (Kemp and Drinkwater 1989). Thus, testosterone could act directly to promote liver tumor induction by binding to the receptor in the target preneoplastic cell. Alternatively, the relevant hormone-receptor interaction could occur in other cells or tissues, resulting in promotion by an endocrine or paracrine mechanism. Because of random X-inactivation, heterozygous (*Tfm*/+) female mice are mosaic for the expression of functional androgen receptors. In order to define the mode of action of testosterone, we treated heterozygous (*Tfm/+*) and wild-type (+/+) B6C3F1 female mice with DEN at 12 days of age and performed ovariectomies at 8 weeks of age (Kemp et al. 1989). Chronic treatment of both heterozygous and homozygous wild-type mice with testosterone resulted in significant increases in the yield of liver tumors relative to control animals receiving only DEN and the ovariectomy. Individual liver tumors were excised and analyzed for androgen receptor activity. As expected, nearly all of the tumors isolated from +/+ mice (35 of 38) expressed androgen receptors at significant levels while approximately one half (12 of 25) of the tumors obtained from *Tfm/+* mice that were not treated with testosterone expressed the receptor. The observation that the proportion (19 of 33) of receptor-positive tumors in *Tfm/+* mice treated with testosterone was similar to that observed for the heterozygotes not receiving the promoter demonstrated that this hormone must act by an indirect mechanism.

The above results are consistent with a model in which testosterone induces the expression of an intermediary secreted growth factor that stimulates preneoplastic hepatocytes to proliferate. The identity and source of this growth signal remain unknown. Androgens regulate the synthesis or secretion of a number of candidate growth factors, including the level of epidermal growth factor (EGF) production by the salivary gland (Barthe et al. 1974) and the periodicity of growth hormone secretion by the pituitary (MacLeod et al. 1991). EGF is a potent hepatocyte mitogen (Richman et al. 1976) and is produced in significantly greater amounts in the salivary glands of male mice than in females (Barthe et al. 1974). We observed that sialoadenectomy resulted in a 40% reduction relative to intact animals in the yield of liver tumors in DEN-treated male mice (C. Kemp and N. Drinkwater, unpublished). However, it is likely that this small decrease in liver tumor induction was an indirect result of a decrease in the caloric intake of the mice that received the surgery (Lagapoulos and Stadler 1987). We are currently testing the hypothesis that growth hormone mediates promotion by testosterone by studying hepatocarcinogenesis in transgenic and mutant mice with altered regulation of growth hormone production.

12.3 Inhibition of Hepatocarcinogenesis in Female Mice

Although chronic treatment of adult mice with some chemical carcinogens results in a greater yield of liver tumors in females than in males
(Innes et al. 1969), female mice are quite resistant to both spontaneous
hepatocarcinogenesis and liver tumor induction by perinatal treatment
with chemical carcinogens (Drinkwater 1989; Kemp and Drinkwater
1989). Comparison of inbred mouse strains for their susceptibilities to
liver tumor induction by perinatal treatment with carcinogen (Kemp and
Drinkwater 1989) reveals that the rank order for sensitivity is similar for
males and females but that the range of susceptibility is reduced for
females relative to that seen for males. Thus, C3H/HeJ male mice are
20- to 50-fold more susceptible to hepatocarcinogenesis than C57BL/6J
males while female C3H/HeJ mice are only threefold more sensitive
than C57BL/6J females (Table 2). The C57BR/cdJ mouse is an exception to this concordance; male mice of this strain are intermediate in
susceptibility, while females are approximately 20-fold more susceptible than females of any other strain (Kemp and Drinkwater 1989;
Poole et al. 1993) (Table 2). We believe that understanding the biological and genetic basis for the unique susceptibility of female C57BR/cdJ
mice to hepatocarcinogenesis will provide us with insights into the
mechanisms underlying the inhibition of liver tumor induction by the
female hormonal environment.

Female C57BR/cdJ mice are genetically insensitive to the inhibition
of hepatocarcinogenesis by ovarian hormones. We treated female
C3H/HeJ, C57BL/6J, and C57BR/cdJ mice at 12 days of age with DEN
(0.05 μmol/g body weight) and subjected groups of mice to ovariectomy or a sham operation at 8 weeks of age (Poole et al. 1993). Intact
C57BR/cdJ mice were 20- and 60-fold more sensitive to liver tumor
induction than C3H/HeJ and C57BL/6J mice, respectively (Table 2).
Ovariectomy resulted in large increases in the yield of liver tumors at
50 weeks of age in C3H/HeJ and C57BL/6J mice; the mean tumor
multiplicities for these two strains were approximately eightfold greater
than those observed for intact mice. In contrast, ovariectomy increased
the yield of liver tumors in C57BR/cdJ mice by only 25%. We can
conclude from these studies that one or more genes carried by
C57BR/cdJ mice circumvent the inhibitory effects of the female hormonal environment observed for other strains of mice.

Table 2. Effect of ovariectomy on hepatocarcinogenesis in mice[a]

	Intact		Ovariectomized	
Strain	Number of mice	Mean liver tumor multiplicity	Number of mice	Mean liver tumor multiplicity
C3H/HeJ	28	1.4	(4.7)[b]	29	11	(7.5)
C57BL/6J	26	0.5	(1.0)	29	4.1	(6.6)
C57BR/cdJ	33	28	(13)	30	35	(14)

[a]Female mice were treated at 12 days of age with N,N-diethylnitrosamine (0.05 μmol/g body weight) and animals were ovariectomized or given a sham operation (Intact) at 8 weeks of age. Liver tumors were enumerated at 50 weeks of age.
[b]Values in the table are mean tumor multiplicity (standard deviation).

Several approaches could be followed in attempting to molecularly clone the genes responsible for the high susceptibility of the C57BR/cdJ mouse. In the candidate gene approach, previously identified genes whose function is consistent with the biology of the phenotype are individually evaluated as to whether or not they determine the phenotype of interest. Alternatively, one could attempt to isolate the gene directly based on differential expression at the RNA or protein level or by DNA transfection with selection for an appropriate phenotype in cultured cells. A critical flaw in all of these approaches is that each requires significant assumptions about the mechanism of action of the gene or the basis, at the DNA sequence level, of the susceptible phenotype. If these assumptions are incorrect, the approach will fail.

In contrast, the "reverse genetic" approach has the virtue of making few assumptions regarding the nature of the gene to be isolated. In this approach, genetic studies of the inheritance of the phenotype are used to determine the chromosomal location of the gene. This positional information is then used to construct detailed genetic and physical maps of the region of the genome containing the gene and DNA sequences contained within the region are evaluated as candidates for the desired locus. This approach has been used successfully to isolate the genes responsible for several human diseases (Monaco et al. 1986; Rommens et al. 1989). Although the first applications of these methods represented extraordinary efforts, recent developments will facilitate greatly the use of a reverse genetic approach to isolate murine cancer susceptibility genes. A critical requirement for approaches based on gene map-

Table 3. Cosegregation of susceptibility to hepatocarcinogenesis in (B6BRF₁ × C57BL/6J) backcross mice with genetic markers on chromosomes 17 and 1[a]

Genotype at		Number of Mice		Mean tumor multiplicity[c]	
D17Mit16	*D1Mit10*				
H[b]	H	10	(18%)	40	(20)
H	B	22	(38%)	16	(18)
B	H	15	(26%)	5.6	(7.0)
B	B	10	(18%)	3.4	(4.8)

[a]Backcross (B6BRF₁ × B6) mice were treated at 12 days of age with diethylnitrosamine (DEN; 0.1 μmol/g body weight) and analyzed for liver tumor development at 50 weeks of age.
[b]H indicates heterozygous for the C57BR/cdJ and C57BL/6J alleles at the marker locus while B indicates homozygosity for the C57BL/6J allele.
[c]Values in the table are mean tumor multiplicity (standard deviation).

ping is the availability of a large number of informative genetic polymorphisms, whose location in the genome is known, for assessment of linkage to the gene(s) that determine the susceptibility phenotype. Within the last 2 years, a genetic map of the mouse based on simple sequence length polymorphisms (SSLP) has emerged as a powerful tool for gene mapping (Dietrich et al. 1992). These genetic markers are based on the frequent occurrence in the mouse (and other vertebrate) genome of simple repetitive sequences, such as the dinucleotide repeat (CA)n. The SSLP markers have several advantages over those based on restriction fragment length polymorphisms or isozymes: (a) they are highly polymorphic among standard inbred mouse strains; (b) they can be analyzed readily by use of the polymerase chain reaction; and (c) the chromosomal locations of more than 3000 SSLP markers have been determined, providing a high density linkage map for the mouse (Copeland et al. 1993). We have used this approach to map the genes responsible for the extreme sensitivity to hepatocarcinogensis of female C57BR/cdJ mice relative to C57BL/6J mice (Poole et al. 1993). For these studies, female mice from a backcross (B6BRF₁ × B6) and an intercross (B6BRF2) were treated at 12 days of age with DEN (0.1 μmol/g body weight). At 50 weeks of age, liver tumors were enumerated as a measure of each animal's susceptibility, and genomic DNA was isolated from the spleen for analysis of SSLP genotypes. The genotypes of each animal at 65 SSLP marker loci were determined, allowing us to analyze approximately 90% of the mouse genome for linkage between the marker loci and susceptibility loci. Inheritance of the C57BR/cdJ alleles for markers on chromosomes 17 and 1 were highly correlated ($p < 0.00005$ and $p < 0.003$) with the susceptibility of female mice to liver tumor induction (Table 3). Backcross mice heterozygous for both loci were more than tenfold more susceptible to hepatocarcinogenesis than mice that were homozygous for the C57BL/6J alleles at both loci. Based on the sensitivities of the mice that were heterozygous for a single locus, the two loci apparently act synergistically to account for virtually all of the high susceptibility phenotype of the C57BR/cdJ mouse. We have designated the chromosome 17 and 1 genes *Flt1* and *Flt2*, respectively, for their effects on "female liver tumors."

Our current efforts are focused on the identification of the *Flt* loci in molecular terms. The chromosomal regions identified in our linkage studies contain two candidate genes of potential relevance to hepatocarcinogenesis. The interval on chromosome 17 that contains the *Flt1* locus also includes the *Igf2r* locus, which encodes the mannose-6-phosphate (insulin-like growth factor II) rececptor (Barlow et al. 1991). This receptor binds and activates members of the TGFβ growth factor family, which act as potent inhibitors of hepatocyte proliferation (McMahon et al. 1986; Dennis and Rifkin 1991). The gene encoding one member of this family (*Tgfβ2*) maps to the same region as *Flt2* (Copeland et al. 1993). We are evaluating the identity between these candidates and the corresponding *Flt* loci in parallel with efforts to identify novel candidates among genes that map to the appropriate regions of chromosomes 17 and 1 that are expressed in the liver.

Given the recent advances in our ability to map genes that control quantitative traits in the mouse, it is certain that, in the near future, the chromosomal locations will be known for a variety of risk-modifier genes that regulate the susceptibility of mice to carcinogenesis in the liver and at other sites. The next step, proceeding from positional information to molecular identification, is a large one, but will be aided by parallel advances in the physical mapping of complex genomes. Ultimately, the challenge will be to relate the function of these genes to human cancer risk.

Acknowledgments. The work cited from my laboratory resulted from the efforts and enthusiasm of several graduate students and postdoctoral fellows, including Dr. L. Michelle Bennett, Dr. Marie Hanigan, Dr. Christopher Kemp, Dr. Gang-Hong Lee, and Therese M. Poole. This work was supported by grants CA22484, CA09135 and CA07175 from the National Cancer Institute (United States Public Health Service).

References

Andervont HB (1950) Studies on the occurrence of spontaneous hepatomas in mice of strains C3H and CBA. J Natl Cancer Inst 11:581–592

Barlow DP, Stoger R, Herrmann BG, Saito K, Schweifer N (1991) The mouse insulin-like growth factor type-2 receptor is imprinted and closely linked to the Tme locus. Nature 349:84–87

Barthe PL, Bullock LP, Mowszowicz I, Bardin CW, Orth DN (1974) Submaxillary gland epidermal growth factor: a sensitive index of biologic androgen activity. Endocrinology 95:1019–1024

Copeland NG, Gilbert DJ, Jenkins NA, Nadeau JH, Eppig JT, Maltais LJ, Miller JC, Dietrich WF, Steen RG, Lincoln SE, Weaver A, Joyce DC, Merchant M, Wessel M, Katz H, Stein LD, Reeve MP, Daly MJ, Dredge RD, Marquis A, Goodman N, Lander ES (1993) Genome maps IV. Science 262:57–66

Dennis PA, Rifkin DB (1991) Cellular activation of latent transforming growth factor b requires binding to the cation-independent mannose-6-phosphate/insulin-like growth factor type II receptor. Proc Natl Acad Sci USA 88:580–584

Dietrich W, Katz H, Lincoln SE, Shin H-S, Friedman J, Dracopoli N, Lander ES (1992) A genetic map of the mouse suitable for typing intraspecific crosses. Genetics 131:423–447

Diwan BA, Rice JM, Ohshima M, Ward JM (1986) Interstrain differences in susceptibility to liver carcinogenesis initiated by N-nitrosodiethylamine and its promotion by phenobarbital in C57BL/6NCr, C3H/HeNCrMTV- and DBA/2NCr mice. Carcinogenesis 7:215–220

Drinkwater NR (1989) Genetic control of hepatocarcinogenesis in inbred mice. In: Colburn NH (ed) Genes and signal transduction in multistage carcinogenesis. Dekker, New York, pp 3–17

Drinkwater NR, Bennett LM (1991) Genetic control of carcinogenesis in experimental animals. In: Ito N (ed) Modification of tumor development in rodents. Karger, Basel, pp 1–20

Friend SH, Bernards R, Rogelj S, Weinberg RA, Rapaport JM, Albert DM, Dryja TP (1986) A human DNA segment with properties of the gene that predisposes to retinoblastoma and osteosarcoma. Nature 323:643–646

Hanigan MH, Kemp CJ, Ginsler JJ, Drinkwater NR. (1988) Rapid growth of preneoplastic lesions in hepatocarcinogen-sensitive C3H/HeJ male mice relative to C57BL/6J male mice. Carcinogenesis 9:885–890

Innes JRM, Ulland BM, Valerio LP, Fishbein L, Hart ER, Pallotta AJ, Bates RR, Falk HL, Gart JJ, Klein IM, Peters J (1969) Bioassay of pesticides and industrial chemicals for tumorigenicity in mice: a preliminary note. J Natl Cancer Inst 42:1101–1114

Kemp CJ, Drinkwater NR. (1989) Genetic variation in liver tumor suscepti-bility, plasma testosterone levels, and androgen receptor binding in six in-bred strains of mice. Cancer Res 49:5044–5047

Kemp CJ, Leary CN, Drinkwater NR (1989) Promotion of murine hepatocarci-nogenesis by testosterone is androgen receptor-dependent but not cell au-tonomous. Proc Natl Acad Sci USA 86:7505–7509

Kinzler KW, Nilbert MC, Su L, Vogelstein B, Bryan TM, Levy DB, Smith KJ, Preisinger AC, Hedge P, McKechnie D, Finniear R, Markham A, Groffen J, Boguski MS, Altschul SF, Horii A, Ando H, Miyoshi Y, Miki Y, Nishi-sho I, Nakamura Y (1991) Identification of FAP locus genes from chromo-some 5q21. Science 253:661–664

Lagapolous L, Stadler R (1987) The influence of food intake on the develop-ment of diethylnitrosamine-induced liver tumors in mice. Carcinogenesis 8:33–37

Lubahn DB, Joseph DR, Sullivan PM, Willard HF, French FS, Wilson EM (1988) Cloning of the human androgen receptor complementary DNA and localization to the X chromosome. Science 240:327–330

MacLeod JN, Pampori NA, Shapiro BH (1991) Sex differences in the ultra-radian pattern of plasma growth hormone concentrations in mice. J Endo-crinol 131:395–399

McMahon JG, Richards WL, del Campo AA, Song MKH, Thorgeirsson SS (1986) Differential effects of transforming growth factor-b on proliferation of normal and malignant rat liver epithelial cells in culture. Cancer Res 46:4655–4671

Monaco AP, Neve RL, Colletti-Feener C, Bertelson, CJ, Kurnit DM, Kunkel LM (1986) Isolation of candidate cDNAs for portions of the Duchenne muscular dystrophy gene. Nature 323:646–650

Moore MR, Drinkwater NR, Miller EC, Miller JA, Pitot HC (1981) Quantita-tive analysis of the time-dependent development of glucose-6-phosphatase-deficient foci in the livers of mice treated neonatally with diethylnitro-samine. Cancer Res 41:1585–1593

Poole TM, Winkler ML, Kaehler DA, Drinkwater NR (1993) C57BR/cdJ mice are genetically insensitive to inhibition of hepatocarcinogenesis by ovarian hormones. Proc Am Assoc Cancer Res 34:255

Richman RA, Claus TH, Pilkis SJ, Friedman DL (1976) Hormonal stimulation of DNA synthesis in primary cultures of rat hepatocytes. Proc Natl Acad Sci USA 73:3589–3593

Rommens JM, Iannuzzi MC, Kere BS, Drumm ML, Melmer G, Dean M, Rozmahel R, Cole JL, Kennedy D, Hidaka N, Zsiga M, Buchwald M, Riordan JR, Tsui LC, Collins FS (1989) Identification of the cystic fibrosis gene: chromosome walking and jumping. Science 245:1059–1065

Rose EA, Glaser T, Jones C, Smith CL, Lewis WH, Call KM, Minden M, Champagne E, Bonetta L, Yeger H, Housman DE (1990) Complete physical map of the WAGR region of 11p13 localizes a candidate Wilm's tumor gene. Cell 60:495–508

Vesselinovitch SD, Itze L, Mihailovich N, Rao KVN (1980) Modifying role of partial hepatectomy and gonadectomy in ethylnitrosourea-induced hepatocarcinogenesis. Cancer Res 40:1538–1542

Vesselinovitch SD, Mihailovich N, Rao KVN, Goldfarb S (1982) Relevance of basophilic foci to promoting effect of sex hormones on hepatocarcinogenesis. In: Hecker E , Fusenig NE, Kunz W, Marks F, Theilmann HW (eds) Carcinogenesis, vol 7. Raven, New York, pp 127–131

Young CYF, Johnson MP, Prescott JL, Tindall DJ (1989) The androgen receptor of the testicular-feminized (Tfm) mouse is smaller than the wild-type receptor. Endocrinology 124:771–775

13 Evaluating Carcinogenic Risks

C. L. Berry

13.1 Preamble

The establishment of a perception that a particular agent is carcinogenic and thus a potential hazard to humans takes place by a number of different routes, both scientific and metaphysical. It takes place against a background in which ill-defined concepts, such as safety, lurk. There is no doubt that the perception of some sections of the community about the cancer-producing potential of chemicals is derived from a pattern of thinking more closely related to the way many defend their articles of faith than to the way in which data are evaluated objectively. Nevertheless, in the selection of hazards for evaluation, public opinion will and should be influential.

If such an evaluation is positive, a risk assessment will be made. This will depend on exposure – the data here will be poor or nonexistent – and on animal studies in which some kind of dose-response relationship will probably have been established. On the basis of this, on assumptions about mechanisms and pharmacokinetics and some free extrapolations, a synthesis will allow risk evaluations to be made from which the acceptibility of any established risk will be determined. The reason for this preamble is to emphasise what sloppy science is going on in this process; the realisation that good and careful experimentation is part of the process is not a justification for the supposition that this A-grade component somehow enhances the rest.

In a thoughtful paper on the use of significance testing in interpreting data Savitz has drawn attention to the misuse of numerical analysis in some biological analyses (Savitz 1993). He points out that the ultimate goal of gathering data is to test some substantive hypothesis about nature; in this context it is clear that the primary operational goal is accurate measurement, not assessing the compatibility of data with an hypothesis (Rothman 1986). This can be done for some of the steps outlined above but in almost all retrospective observational epidemiological studies of carcinogenicity there is no randomisation; those observed underwent particular exposures in a far from randomised manner and there is thus no proper basis for the use of p values which properly can only be utilised in answering the question "If the null hypothesis were true and an infinite number of replications of the experiment were performed by drawing new, random samples, what proportion of those replications would yield results as or more extreme as those obtained" (Savitz 1993). This stringent requirement may be applied in human epidemiology but is seldom used in the extrapolation of data of spurious precision to provide estimates of population tumour risks. So we see a further justification for the note of caution expressed by Purchase (1987) from different premises "we find ourselves in the curious situation of using data from epidemiological studies ... to try to disprove a hypothesis which is based on a sequence of untestable assumptions".

I would like to make one further biological point at this time; much of the discussion of the problem ignores the data base on the species we seek to protect, humans. Age-corrected rates for cancer are not changing except where well-established exposures to the few human carcinogens we have identified are themselves changing; these changes are not always for the worse (lung cancer rates in UK males continue to fall). Occupational health data continue to improve but until better measures of exposure are devised epidemiology will continue to be a flawed method of risk assessment. But this may not be the only way forward; we have heard a great deal about mechanisms of carcinogenicity in the last 2 days and it is a matter of great excitement that much of this newly derived information is already influencing medical practice. Detailed accounts of the genetic changes in colon cancer, a major cause of death in most developed countries, have led so much better an understanding of the chronology of events as to give rise to proposals for prevention by once-only sigmoidoscopy at 55–60 years with subseqent surveillance for the 3%–5% of individuals found to have high risk lesions (Atkin et al. 1993). The savings over the currently recommended faecal blood/endoscopy regime every 3–5 years would even please Mrs. Clinton, but the knowledge base on which they depend should certainly influence regulators.

13.2 Models

And so we come to models. The most recent Eurotox Newsletter (2/93) reported a debate on the regulation of dioxins. In the report it was concluded that the differences between the US and European approaches might be summarised as a dependance on the application of mathematical models (mainly the linearised multistage model) in the US, and a classical 'threshold with safety factor' approach in Europe. A range of several orders of magnitude of acceptable exposures might be arrived at using the two models and the same data and one of the discussants (Dr. Birnbaum) pointed out that a number of "fairly compelling arguments" can be deployed in the support or rejection of either method. If it is necessary to allay public anxiety about chemical exposure (and I am certain that it is) then this type of sophistry is counterproductive – it should be engaged only in a contex that makes plain that what is being discussed is management, not science. Let us talk about models in this context.

All models depend on a number of assumptions:

1. That the results of animal experiments can be extrapolated to humans.
2. That the dose response is linear.
3. That preferred models of carcinogenesis are applicable to human tumours in a generalised manner; concepts which include different routes to the end-point of uncoordinated cell proliferation or to different numbers of steps in multistage processes are not considered.

In the EPA's q* value approach the 95% upper confidence limit on the slope of an (assumed linear) dose–response line is used to provide unverifiable conservative data which are insensitive to the biological data. Purchase has illustrated problems with this methodology and has shown that using the TD50 (the dose rate in milligrams per kilogram per day which, if admininstered chronically for a standard period, would halve the probability of an animal remaining tumourless; Peto et al. 1984) and accepting linearity in dose response allows the relationship between the 50% tumour-inducing dose ($qx5 \times 10^{-1}$) and the dose to produce a 10^{-6} risk ($qx10^{-6}$) to be determined as the constant, 5×10^{-5}. He has shown that this is a valid approach for 38 carcinogens. This has the intellectual honesty of being a transparent manipulation of data with no additional science inferred; it is flawed (because of the assumptions) but intellectually honest. Calculating the "virtually safe dose" – what a phrase! – is a matter of division.

It is possible to make very conservative assumptions to produce data which result in enormous estimates of risk but my dislike is not really the selection of assumptions but the whole list; all of them are flawed.

13.3 Particular Problems

Hay (1991) has identified the difficulties with maximum tolerated dose testing in rodent cancer bioassays, in particular the problem of tissue injury with the induction of cell division. The decreased interdivision interval which is a consequence of this activity "fixes" DNA injury by allowing inadequate time for repair, and we have a clear indication from that this is an important mechanism in human carcinogenesis. p53, a cell cycle-related DNA binding protein which regulates transcription, acts as a tumour suppressor gene and in many human tumours one allele is mutant and the other deleted (Weinberg 1991). The gene product acts as a cell cycle check point leading to S-phase delay in cells with genetic damage, thus allowing more time for DNA repair before changes are fixed by the next cell division (Lane 1992). Levels of p53 appear to be regulated by differential gene expression as well as at the post-transcriptional level and there is evidence for a particular role for this gene in differentiation; it has been shown that its level of expression in fetal mice is not well correlated with cell proliferation, but upregulation of p53 gene expression may be necessary to inhibit cell cycle progression and to allow terminal differentiation at many sites (see Lane 1992). These data, together with an increasing volume of information on the role of growth factors and oncogenes acting as highly conserved modulators of the differentiated state in all of the Animalia (see Berry 1994) make clear an important failure of all models; none consider DNA repair in an adequate way. I am sure that we have all heard Bruce Ames' comment on the number of oxidative hits per cell per day and even conservative estimates provide values of a least a thousand. Indeed, Lindahl (1993) provides data suggesting that 100–500 hits/day occur in human cells – hydroxyl radicals generate 8-hydroxyguanine which base-pairs preferentially with adenine rather than cytosine, producing transverse mutations after replication. However, as Lindahl has pointed out spontaneous damage to DNA occurs by other routes and probably at greater rates. Hydrolysis is important, DNA in cells depurinates at the same rate as DNA in solution and between 2000 and 10 000 DNA pyrine bases turn over per day – apurinic endonucleases have a repair capacity which comfortably exceeds spontaneous hydrolysis rates.

13.4 A Proper Use of Models

Let us take a completely different field as an example of what can be done usefully with a model. In the formation of the primary germ layers, stochastic rather than programmed events cause most of the changes. The ventral furrow, posterior midgut invagination and other morphogenetic movements in *Drosophila* occur where the regular cobblestone like appearance of the ectoderm is disturbed by rows of cells which undergo apical constriction, forming a groove (Sweeton et al. 1991). Nuclear positions in the cells in the groove change following cytoskeletal changes and eventually, due to the process of cell shortening which appears to occur in a stochastic manner, accumulated changes in cell form result in invagination. The groove is produced not by a grand genetic plan but rather by local changes in the cellular environment, which at a given stage in the genealogy affect the cytoskeleton. Presumably using this type of information as a stimulus Clausi and Brodland (1993) have carried out some very convincing computer simulations of the process of neurulation. The process of neural tube closure has been of major interest in embryology and dysmorphogenesis and is clearly dependent on changes of the kind described in the work of Sweeton et al. (1991) above, but a great many variables have been shown to influence the process. Their model allows for the nonisometric properties of tissue and cells and for the large strain values exhibited by some biological materials in the construction of a "virtual" embryo. They have shown that when uniform isotropic circumferential microfilament bundle (CMB) constriction and chephalocaudal elongation act together on a simulated circular neural plate it becomes keyhole shaped, resembling the neural plate, and that when these forces act on a spherical (amphibian) embryo dorsal surface, flattening occurs. CMB constriction can produce sequential formation of neural ridges, narrowing and thickening of the neural plate and neural tube closure. This series of events is produced by mechanical changes alone; no cellar divisions or nonmechanical cell–cell interactions are necessary.

Most importantly, multiple "redundant" mechanisms are necessary to produce an effective result; it is interesting to note the comments of Wolpert (1992) in this context. He has pointed out that some cases of redundancy may be evolutionary relics where genes have persisted for millions of years because they present no selection disadvantage; others appear to be necessary. In general it is true to say that many growth control and cell signalling mechanisms are used repeatedly in development, often in different ways at different stages (see Berry 1992, 1994). These various gene products are almost entirely identical with the factors disturbed in carcinogensis. An important characteristic of developmental mechanisms is that there often appears to be considerable redundancy in the operating systems – incidentally, a point pathologists should recognise when they identify particular gene products as "characteristic" of particular cell types. In a number of cases of apparent redundancy, the additional mechanism may be effective in increasing the specificity of a process, acting as a form of "fine tuning" of a regulatory process – an analogy would be the differing processes which contribute to the effective monitoring of the fidelity of DNA reproduction. An example of this tuning is seen in the development of *Drosophila* where genes determining anterior and posterior characteristics of an embryo interact with genes having localised nonpolar expression, to produce graded concentrations of gene products which may have activating or repressive effects.

Now, in this kind of model neurulation provides a well-defined end point; "a neoplasm" is an indeterminate definition, clearly nonsensical in terms of pathogenesis. The model depends on a mechanical property which is measured and which can be applied in a framework of physical laws which are well established universally – despite the wishes of many biologists the laws of the physical universe apply to biological materials.

Even in this (by our sloppy standards) rigorously defined system, multiple redundant mechanisms are necessary to obtain an optimal effect. We are a long way from this in carcinogenicity.

13.5 Conclusions

It should be made clear that numerical models of carcinogenicity are devices which are used to facilitate the administrative difficulties of regulation of hazardous chemicals. They are constructed in a manner which ensures an exaggeration of possible risks and this may well be the political option to be preferred. They should not be presented to the public in a manner which suggests that they are part of the scientific paradigm; this is false, and, more importantly, damages the scientific credibility of toxicology.

References

Atkin WS, Cuzick J, Northover JMA, Whynes DK (1993) Prevention of colorectal cancer by once-only sigmoidoscopy. Lancet 341:736–740

Berry CL (1992) What's in a homeobox? The development of pattern during embryonic growth. Virchows Arch [A] 420:291–295

Berry CL (1994) Building an embryo with limited resources. In: Barness LA (ed) Advances in pediatrics, vol 41. Mosby, St. Louis

Clausi DA, Brodland GW (1993) Mechanical evaluation of theories of neurulation using computer simulations. Development 118:1013–1023

Hay A (1991) Carcinognesis: testing time for the tests. Nature 350:555–556

Lane DP (1992) p53, guardian of the genome. Nature 358:15–16

Lindahl T (1993) Instability and decay of the primary structure of DNA. Nature 362:709–715

Peto R, Pike MC, Bernstein L, Gold LS, Ames BN (1984) The TD50: a proposed general convention for the numerical description of the carcinogenic potency of chemicals in chronic-exposure animal experiments. Environ Health Perspect 58:1–8

Purchase IFH (1987) Carcinogenic risk assessment: are animals good surrogates for man. In: Bannash P (ed) Cancer risks, strategies for elimination. Springer, Berlin Heidelberg New York, pp 65–79

Rothman KJ (1986) Modern epidemiology. Little and Brown, Boston

Savitz DA (1993) Is statistical significance testing useful in interpreting data? Reprod Toxicol 7:95–100

Sweeton D, Parks S, Costa M, Weischaus E (1991) Gastrulation in Drosophila: the formation of the ventral furrow and posterior mid-gut invaginations. Development 112:775–789

Weinberg R (1991) Tumor suppressor genes. Science 254:1138–1146

Wolpert L (1992) Gastrulation and the evolution of development. Dev Gastrul [Suppl] 7–13

Subject Index